THE POLITICS
OF
AGRICULTURAL
RESEARCH

The Politics
of
Agricultural
Research

DON F. HADWIGER

WITHDRAWN

UNIVERSITY OF NEBRASKA PRESS

Lincoln and London

The paper in this book meets the guidelines for permanence
and durability of the Committee on Production Guidelines for
Book Longevity of the Council on Library Resources.

Library of Congress Cataloging in Publication Data

Hadwiger, Don F.
The politics of agricultural research.

Bibliography: p.
Includes index.
1. Agricultural research—United States.
2. Agricultural research—Political aspects—United States.
3. Agricultural research—Government policy—
United States. I. Title.
S541.H25   363   81-24077
ISBN 0-8032-2322-6   AACR2

CONTENTS

PREFACE

Agricultural research institutions, after many years of operating in relative obscurity, are "coming out" into the world they have changed so much. It is an appropriate time for them to do so.

Public attention to agriculture has generally focused upon the farmer as producer, with little notice being taken of the fact that the farmer has had many sets of elves. Thus the work of those inconspicuous research institutions which have done much of the training, norm-setting, thinking, and communicating—and even deciding and implementing—in the agricultural world has usually been ignored or overlooked.

Now, however, a large nonagricultural public has become interested in agricultural research. Meanwhile, the agricultural subsystem has chosen to limit its involvement in research activities and no longer provides its research institutions with adequate support or adequate scope. Therefore, these mature and quite successful institutions are being pressed to change their associations and approaches, and to rewrite their job descriptions.

For the institutions and for those interested in their work, this is a moment of anticipation, a time for questions to be raised and decisions made about the future of agricultural research. On the assumption that, if each of those involved in the decision making can see where the others have been and where they want to go, all can move with more grace and effectiveness, indeed with less pain, this book attempts to describe the present state and goals of agricultural research. The book is about political demands and political supports, and about the political processes that lead to political decisions. It addresses the questions "Who makes agricultural research policy?" and "What do they want it to achieve?" The book may offer, as well, some insight into the impact of agricultural research, though that subject is treated only indirectly.

Other important elements of the agricultural scene are touched upon only as they relate to U.S. public agricultural research institutions. These include research undertaken by private industry and research sponsored by private foundations and international research centers, as well as that conducted by other federal research agencies such as the National Institutes of Health and the National Science Foundation. The cooperative extension services, which are linked with agricultural research institutions in their histories, in their budgets, and in their policy processes, are discussed in those contexts and in some case studies. There is virtually no discussion of other nations' research, although readers should be aware that there are distinguished and productive agricultural research institutions elsewhere in the world. It is to be hoped that readers will agree that the politics of U.S. agricultural research is a subject of sufficient scope and importance to warrant the intensive treatment this book offers.

ACKNOWLEDGMENTS

Primary and secondary sources for this book are discussed in the bibliography. Formal interviews were conducted with scientists, administrators, interest group leaders, and others involved in agricultural research policy. I am grateful for all these sources, and especially grateful to those who gave their time in interviews.

Perspective was gained from several work opportunities for which I am indebted. I served for one year in the USDA's Economic Research Service (1965–66), for six months in the Agricultural Research Service (1977), and for many years after 1962 as a faculty member at Iowa State University, which is a land-grant institution. For a full year, in 1969–70, I was employed by the Washington Research Project, which advocated most effectively the rights of poor people in the rural South and elsewhere. Much earlier, in 1959, I served for nine months as an American Political Science Association Congressional Fellow with the staff of the U.S. House Agriculture Committee and in the office of Senator John Kennedy.

Several scholars read earlier versions of this manuscript. Albert Grable, Garth Youngberg, Charles Gratto, Jerry Stockdale, James Hildreth, and Vivian Wiser made many good suggestions which were incorporated into revisions of the manuscript; for the many shortcomings which no doubt remain, I take full responsibility. Besides praising these generous readers, I thank them warmly.

A number of persons provided research assistance during the preparation of this manuscript, including Paul Gardner, David Ekhomu, Dennis Yergler, Justin Halperin, Robert Boos, Janet Gardner, David Hadwiger, and John Greenwood. Secretarial assistance was provided during several manuscript revisions by Risë Pennell, Sherry Smay, Kathy Fyock, Edna Henry, Nancy Olson, and Jane Windom. I am most grateful for all of this excellent assistance.

Acknowledgments

Finally, I wish to express appreciation to the Iowa Agricultural Experiment Station, the Agricultural Research Service, and the Department of Political Science at Iowa State University for their substantial research support, and in the case of the last two named, for providing me a stimulating work environment while the manuscript was in preparation.

CHAPTER ONE

Introduction

The impact of agricultural research is displayed on every hand, unlike that of social programs, whose results are difficult to measure. In the supermarket, as a striking example, there is the wall of handsome fresh vegetables, most of them available throughout the year. In the United States, food costs less as a percentage of income than elsewhere in the world, which cannot be said of such other essentials as medical care, transportation, and education.

Other results of agricultural research may not be so praiseworthy. On an international standard, some supermarket foods may fail a taste test, taste having sometimes been sacrificed to convenience in handling and manufacture. Our diets are probably not as healthful as they might have been, yet are better than we deserve, given the neglect of nutrition research.

Out in the country enormous farm implements parked alongside decaying barns give a hint that most of the farmers are gone and most of the farmtown mainstreets are dead. Did the millions who left, including 700,000 sharecroppers, go on to a better life? In the developing countries of the world, our labor-saving technology has pushed people off the land, and thereby "wasted" the labor which is their most valuable resource.

Other results of agricultural research can be seen, felt, and even smelled. Texas cattle produce more organic waste than do Texas citizens, and some of this waste blows or washes from the land. Soil erosion increases when terraces are removed to accommodate giant machinery. Other soil-conserving practices are abandoned under specialized cropping. Research scientists are pushed by their clients to find ways to adjust to soil losses, rather than to prevent them. Tourists on Western highways can spot the areas already blighted by salty irrigation water, for which researchers are seeking saline-tolerant plants.

Does agricultural research help man to prepare for the future, or does it encourage him to ignore it? Even as we scavenge the world's dwindling resources, human population is on the way to being doubled within a few decades. Will agricultural research provide the miracles we need to feed and employ these people? One can argue that we should quit depending upon the mixed blessing of new technology, and simply use what we have more equitably, as in China. But one can also argue that we should coax new science miracles that will restore our soil, that will trap for us more of the sun's abundant energy, that will allow others in the world to share our leisure and abundance, rather than obliging us all to readapt to manual labor and Spartan diets.

It *must* be possible to innovate even while sharing and conserving. This has been the assumption of three country boys who are exemplary of many, many others, each contributing in more or less spectacular ways to the explosion and management of food technology. Ray Jackson is a scientist whose discoveries made the cover of *Science Magazine*. Sylvan Wittwer is a research administrator who made the case for more basic research. Jim Hightower is a journalist whose exposé helped bring about a reappraisal of the marvelous but flawed product of agricultural research.

RAY JACKSON

The problems—and the possibilities—of agricultural research are exemplified in Ray Jackson's work. Dr. Jackson stands in his three-acre wheat field, a photo of which appeared on the cover of *Science Magazine*, pointing his Model 44 Derringer thermometer at a nearby patch of cotton. The thermometer gun shoots infrared rays, the rebound from which reports the temperature of the cotton plants.

Much is learned by taking temperatures. "The best way to find out if a plant is sick is to ask it," Jackson says. "A plant that is hotter than the air around it is experiencing stress. It's not getting enough water, or it may be under attack."

The air temperature is above 100 degrees, because this is a summer afternoon in Phoenix. Jackson's laboratory was placed here in 1953, at the wish of a senior member of the Senate Appropriations Committee, Carl Hayden of Arizona. Jackson has stayed at this desert job for many years because Phoenix has become "home" to him, his Mexican-American wife, and their family, and because he feels grateful to the federal government, which through the GI bill and a part-time job, enabled him to obtain a Ph.D. degree, and then sponsored him in a full-time science career: "Every time I achieve something, I feel like I'm paying them back a little bit." Now, as a productive scientist, he is given whatever he needs to pursue his research leads, including an able team of scientists. "If Ray Jackson says he wants an entomologist, give him one" is the attitude of his Washington supervisors.

Jackson's group helps set new directions for agriculture research through its basic findings. "I have found that Washington is responsive to us in demanding a follow-up on our results," says the team's theoretician, Sherwood Idso.

The Phoenix laboratory is a "four-star" contributor to the ongoing technological revolution in agriculture, one of whose spectacular manifestations is the train of busy machines that crosses a commercial farm. Nowadays the farmer may sit in a lofty, air-conditioned, glass-walled cab, working a bank of levers and buttons as he listens to music on his radio. Behind him is a variety of farm implements, each sporting unique combinations of hoses, wheels, discs, and tanks, hissing chemicals into the soil while also stirring and seeding it, as they sweep across a dozen former homesteads.

This ultimate farmer may himself be displaced by a laser beam. As developed by scientists in the Phoenix laboratory, the laser device already guides machines in leveling land for irrigation, and may soon guide tractors around a field.

Human surveillance is removed still another step by "remote sensing," another contribution of the Jackson-Idso team. The group demonstrated, after years of monitoring, that soil temperature can reveal soil moisture levels and that plant temperatures reveal plant stress. The team helped develop means for infrared monitoring of temperatures from car windows, airplanes, and satellites.

The implications set minds swimming. Crop failures may be predicted and famines prevented by satellite monitoring of crops throughout the world. National and international agencies are now attempting to accomplish this. Remote sensing can provide early warning of insect threats; information about pest vulnerabilities may be used to frustrate the pests more effectively than the blankets of insecticides which nervous farmers now lay down. In the water-scarce West, remote sensing can more precisely ration irrigation water.

Says Ray Jackson: "One thing I dream about is a tethered satellite, which would stand above us at about one hundred thousand feet—out of the weather winds and with the energy to keep it in place. It would have a multitude of sensors—microwave and infrared—and each farmer—even the small farmer—would have a device not much bigger than a pocket radio through which he would be able to interrogate the satellite about water conditions, insect conditions, all the things that affect his plants."

This scientist's dream is not science fiction, but is it social fiction? Jackson's remote sensor will probably be less valuable to the small farmer who can walk his fields each morning than to the big city land firm which controls thousands of unobserved acres.

American agriculture has become big agriculture, in part because larger farms and big agribusiness have been the chief beneficiaries of agricultural research, but for other reasons as well. The research establishment did not resist the wave of bigness which had earlier swept across nonagricultural sectors. Indeed, agricultural research was on the leading edge of this wave—"in the groove," as surfers say.

The results of the application of research findings to innovative, capital-intensive, inevitably "big" farming operations have been dramatic. Farm output per unit of input more than doubled between 1910 and 1975.[1] There was a threefold increase in gross output of food grains, because of greater per-acre yields. U.S. agricultural growth kept ahead of population growth, in contrast with most other countries, with the result that agricultural products have become a major export item, helping to offset the drain of oil imports.

Much of the credit for this achievement should go to agricultural research. The return from research, as a percentage of public funds invested in it, has been in the range of 30 to 60 percent per year.[2] On the basis of this uncommonly high return, one would expect much more to be spent for research, and no doubt many economists have scratched their heads over "society's" failure to do so. Economists would have been puzzled in the nineteenth century also, because "society" then invested money in agricultural research even though it yielded virtually nothing of value.

Has there been any growth in agricultural research funding? In constant dollars, funding for the state research institutions doubled between 1950 and 1970.[3] But national-level funding has remained stable even as funding for other federal agencies has increased.[4] Overall, the number of agricultural scientists funded by government has declined.

Why the lack of growth? Centers of resistance to bigger budgets, until 1981, were the president's Office of Management and Budget and the office of the secretary of agriculture. Economist Vernon Ruttan, seeking reasons for their resistance, suggests that, as a matter of policy, the government wants to avoid food surpluses, and therefore does not want to increase productivity beyond the rate of growth in demand for food products.[5] But this explanation credits the policy makers with looking beyond the next election, since changes in research funding are unlikely soon to affect production. In another sense it paints them as dunderheads, ready to hobble a major export industry simply because a few experts predict a middle-range world surplus. (In some quarters, the belief is fondly held that OMB, in particular, is a stuffed-with-numbers group of know-nothings.)

Perhaps the true dunderheads are the consumers and other "diffuse interests" who benefit from agricultural research but do not support it. Is it possible to mobilize support from this diffuse constituency?

Some research clients have been mobilized, as we shall see. Research institutions have developed close, even intimate, ties with some of those who produce and market our food. A bond of mutual interest now unites research leaders with farm and agribusiness leaders, and with the legislators whose districts rely upon agriculture. These leaders may also share a cultural background of farm upbringing and agricultural college training (although most rural legislators are lawyers). There is, in any case, continuing interaction among them; indeed, they have long been organized into an industry-wide political coalition.

For many decades, this group was privileged to determine U.S. agricultural policy in all its aspects, including economic structure, prices, production, grades, research, and rural social policy. The coalition was envied for its size, its aggressive leadership, its effective use of "down-home" imagery, and its successes both in economic productivity and political influence. The farm coalition created an enormous body of regulatory law in its own behalf and commanded large governmental subsidies, seeking to use them exclusively and arbitrarily in behalf of coalition members.

Farmers today are less important to the coalition, whether as voters or as contributors to the agricultural industry. But there are still centers of power, such as Congressman Jamie Whitten's agricultural appropriations subcommittee, which seek to maintain the agricultural coalition. They are burdened by a dispiriting fragmentation of commodity interests, ideologies, political parties, regions, and farm organizations. Even the various agricultural research institutions alternate between hesitant cooperation and petty bickering among themselves. In order for agricultural interests to command a congressional majority on major farm bills, it has been necessary to form a broader-based coalition, one that usually includes labor and welfare interests.

Meanwhile, a counterforce to the agricultural coalition has emerged, challenging its formerly bigoted attitude toward minorities, its thoroughgoing exploitation of farm workers, and its cavalier indifference to consumer nutrition and health, among other things. In the 1960s the critics of agriculture included environmentalist Rachel Carson and the Hunger Lobby, which succeeded in diverting agricultural funds to a food stamp program for the poor. In the 1970s, Jim Hightower led a direct attack upon the research institutions, and other critics joined in.

Paradoxically, the self-appointed critics of agricultural research were in the same tradition as some of the self-appointed public interest advocates of the nineteenth century who brought these research institutions into being. The earlier activists included editors, agriculturalists, rural aristocrats, civil servants, and enthusiastic young scientists. With much effort, these nineteenth century reformers convinced federal and state governments that public funds should be spent for agricultural research as a way to improve the lives of the nation's farmers.

The modern public interest advocates, by comparison, have wished to reshape agricultural research by exposing its shortcomings, by regulating it, and by committing its funds to previously neglected areas such as human nutrition. In anger, and perhaps unwisely, the research establishment at first rejected these critics, who were, as it turned out, offering to research institutions the prospect of new missions and new political support.

JIM HIGHTOWER

One of the new critics, Jim Hightower, a "98-pound weakling" with steel-rimmed glasses on a small face and a crown of red hair spreading like a cape over his shoulders, became a raging prophet who made a public issue of the question of whether public agricultural research was indeed serving the public (particularly the rural public) or whether it was performing a disservice to that public.

Hightower came and went quickly as a radical critic of agricultural research, publishing his critique of land-grant colleges at age twenty-nine (*Hard Tomatoes, Hard Times*)[6] and three years later publishing an exposé of the food industry (*Eat Your Heart Out*),[7] then moving on to other activities.

Proud of having been reared in a Texas backcountry setting, and educated at an undistinguished college without benefit of any elite credentials, Jim Hightower regarded himself as a rural populist. He observed that corporations and other large entities used public institutions for their own ends, often with disastrous consequences to rural people and communities. Agricultural research and extension services within the state land-grant colleges became his principal case in point. Hightower's thesis in *Hard Tomatoes, Hard Times*, repeated in symphonic variations, was that the agricultural colleges had been created to serve the common people but had abandoned them to serve an "industrial elite."

He quoted establishment leaders who admitted that research and extension did little for small farmers. He listed twenty-five commodities for which researchers were seeking to develop mechanical harvesters, presumably to escape dependence upon farm workers. As evidence that research leaders were in alliance with big business, he printed the biographical statement of then-Secretary of Agriculture Earl Butz, who while dean of the School of Agriculture at Purdue University, had simultaneously served on the boards of three giant corporations, one each in animal feed, agricultural chemicals, and farm machinery. After presenting such evidence, Hightower described its consequences: "The great majority of rural Americans are strangers to these public laboratories that were created to serve them. When research does not ignore them, chances are it will work against them. If they do get help, it comes either in the form of a meager trickle that has been carefully sluiced and strained upstream, or in the form of irrelevant and demeaning sociological probes into their personal habits."[8] Furthermore, he noted: "While rural towns shrivel and megalopolis becomes hopelessly congested, land grant researchers are using tax dollars to concoct managerial schemes and to design technological systems that will send millions more packing off to the cities. Tax dollars buy new tinker toys for agribusiness, misery for migrants, death for rural America and more taxes for urban America. All in the name of efficiency."[9]

When USDA official Don Paarlberg, who shared some of Hightower's concerns, said his book was "too provocative", Hightower explained:

> You have to be to get any movement. Agriculture is a very complicated subject, and it does not easily arouse the attention of urban people or urban congressmen. For years, urban congressmen have let Earl Butz and a few agriculture state congressmen write agriculture policy. The result has been a disaster for family farmers, consumers, taxpayers, and workers. We write our reports in a way that these congressmen can see their own self-interest in agricultural policy. That is our role—provocateurs. There are plenty of "responsible" groups. What we try to do is put issues on the front pages of newspapers and create a climate for change. Then the other groups can go from there.[10]

Paarlberg later coined the term "new agenda" to suggest that the concerns of commercial agriculture had been replaced by the heretofore neglected concerns of and for small farmers, farm workers, the poor, racial minorities, consumers, and the environment. Paarlberg attributed this shift in emphasis to the efforts of Hightower and of other new groups of public interest advocates.

SYLVAN WITTWER

Still another research agenda was being created by some aggressive and concerned scientists who organized scientific panels under such prestigious sponsorship as that of the National Academy of Sciences and began to lobby legislators and administrators. The panels generally concluded that the world's agricultural systems must be modified in order to meet the next century's food needs.

Predictably, these scientists envisioned a major role for research. They advocated basic research, seeking breakthroughs in fundamental knowledge through which agriculture could be made to give man more food for fewer inputs.

One of the scientists involved in the effort was Sylvan Wittwer, director of the Michigan State Experiment Station and a flamboyant member of a broader natural science elite.

According to Wittwer, speaking in 1975, prospects were good not only for major breakthroughs in food technology comparable in scale to the development of the atomic bomb, but also for multiple small discoveries whose cumulative effect would be equally significant. But this would happen only if the government were willing to support food research with the same intensity that it had shown for the Manhattan Project which produced the atomic bomb.

"Do you realize," Wittwer exclaimed to the House Science and Technology Committee,

> that the key to practically all food production on this Earth and the efficiency of the photosynthetic process is due to a single enzyme, and man's ability to control the functioning of that enzyme, namely ribulose diphosphate carboxylase. Furthermore, that it is this one enzyme that makes the difference between plants, such as corn, maize, and sugar cane and sorghum, which we, on the one hand, call the C-4 plants, and which have a remarkable ability to efficiently utilize the energy of the sun, as compared to the much less efficient soybean, cotton, potato, and the small grains, which we call the C-3 plants? There must be means by which the action of this particular enzyme, this key enzyme, could be regulated chemically, genetically, or both.
>
> There are less than a dozen laboratories in the United States that are actively engaged in research in this area . . . there could be 10, 20, 30, 40, 50, 100 times more funds used in this area, and used profitably.[11]

Wittwer had many other suggestions for "Manhattan Project" priorities, which were also stated as high priorities in reports by scientific panels: nitrogen fixation, a process by which the plant fertilizes the soil; converting plant and animal wastes into creation of energy; integrated pest management, featuring biological attacks on insects; allelopathy, creating plants "which have their own built-in mechanism to inhibit the growth of weeds"; more effective ways to use ruminants, that is, to use cows, sheep and goats to digest the plant cellulose that humans cannot digest ("ruminant animals need not compete with man for sources of food"); improving environmental quality by multiple cropping and reduced or zero tillage, protected cultivation, and controlled environment agriculture.

Professor Wittwer was not without ideas for changing consumer habits. "The dietary and health aspect of predominantly vegetarian diets should be reviewed. If we are to provide food for other nations, we should ask ourselves: Feed the world with what? Certainly not the diet we eat. If not with what we have, then what? Then the moral issue of double dietary standards comes to the front."

The scientists and the public interest advocates were persuasive. In the 1977 Food and Agriculture Act, new agendas covering past omissions and future challenges were united with the continuing agenda of servicing U.S. commercial agriculture. The law authorized a doubling of federal funds for agricultural research. In Congress, a grand coalition approved this new research charter almost unanimously.

Despite the apparently universal enthusiasm for the new program, difficulties arose in its implementation. Those who passed the new agenda lacked confidence that the research establishment would adopt it in practice, and indeed the research institutions themselves seemed reluctant to take on new functions.

Thus, the search continues for an organization of research which can address the new and old agendas, and thereby secure the much expanded financial support needed to bring both to fruition.

CHAPTER TWO

The Shaping of Agricultural Research Institutions

Although research institutions have produced a highly integrated technology and have been important in developing a social subsystem based on agriculture, these institutions are decentralized in structure.

There are agricultural research institutions at both national and state levels (fig. 2.1), as one might expect in our federal system. Within each of the states is at least one land-grant university, so named because it was supported during infancy by a federal grant of land. Among the major subdivisions of each land-grant university are colleges of agriculture and home economics, in which a number of academic departments are organized according to specializations such as "Agronomy" and "Textiles and Clothing." Most agricultural scientists are located within these academic departments. Also incorporated within the university structure is an administrative entity called an agricultural experiment station, which channels research funds to each academic department. These research funds are obtained partly from federal grants distributed to the states on the basis of formulas which, for example, provide larger grants to states with relatively large numbers of rural citizens.

In the typical state, federal "formula" funds comprise less than

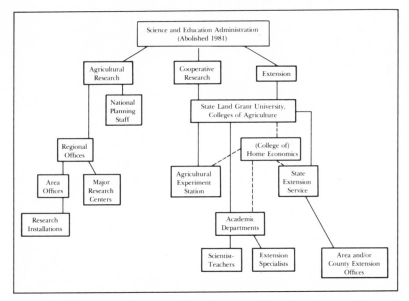

FIGURE. 2.1: Organization of Agricultural Research Institutions (Organization varies somewhat by state. There are major variations in the positioning of Home Economics.)

one-fourth of the total experiment station budget. Most funding is obtained from the state legislature.

Another agency within the university is the state cooperative extension service, which supports a staff of professionals both within the university's academic departments and in county and regional field offices. It is extension's job to communicate research findings to citizens. The cooperative extension budget is also provided from a combination of federal and state appropriations.

The directors of the agricultural experiment station and the cooperative extension service are answerable to the dean of the college of agriculture, who may himself hold all three titles. The dean is appointed by the president of the university; but the experiment station and extension service sometimes function almost as independent entities. Considerable discretion is allowed also to the chairpersons of the academic departments and to the research professors themselves.

There are national offices for the state extension services and experiment stations; these national agencies serve mainly in a supportive and coordinating role, but they may also urge attention to national research priorities.

The colleges of agriculture, in which research and extension are conducted, also have the function of training future leaders— including scientists—from the time of their entry into grade school 4-H clubs until they finish professional training. The colleges also have served as a crossroads meeting place for agricultural leaders. As centers of training, research, and communication, these colleges have helped produce a social subsystem for agriculture.

At the federal level, too, there are agencies which specialize in agricultural research. There were a number of federal agricultural research bureaus until 1952, when most of these were merged into the Agricultural Research Service (ARS). Then in 1977 the ARS itself was folded into a national superagency called the Science and Education Administration (SEA), which oversaw state research and extension as well as the National Agricultural Library. In 1981, SEA was abolished, and its components regained agency status: the Agricultural Research Service, the Office of Extension, the Cooperative States Research Service, and the National Agricultural Library.

Social science research at the federal level was conducted by the Economic Research Service until this agency was recently merged with two other federal agencies: the Statistical Reporting Service, which predicts crop outputs and collects other data; and the Farmer Cooperative Service, a small agency serving cooperatives, which has since been removed. The combined agency was called the Economics Statistics Cooperatives Service. In 1981, this superagency too was abolished, and its components were returned to agency status.

The federal research agencies not only work for the agricultural industry, but also serve the agricultural research needs of federal "action" agencies, such as the USDA's Soil Conservation Service. But one powerful "action" agency, the USDA's Forest Service, has maintained its own large research division, and therefore should be regarded as an additional federal research agency.

The federal and state agricultural research institutions are the principal subjects of this book. We should keep in mind that other public agencies support some agriculture-related research, including the National Science Foundation, the National Institutes of Health, the Department of Defense, and several international agricultural research centers financed in large part by the U.S. Agency for International Development.

HISTORY OF STATE & FEDERAL
RESEARCH INSTITUTIONS

The history of agricultural research policy is one of federal and state governments playing leapfrog. First, during the early nineteenth century, a few states instituted agricultural experiment stations. Then, in 1862, Congress created the Department of Agriculture (USDA) as a research institution which would gather and analyze information for farmers, and in the following decades federal research bureaus grew up with the USDA. Also in 1862 Congress passed the Morrill Act, a federal grant-in-aid to the states, which encouraged the states to sponsor colleges that would teach agricultural and mechanical arts. In 1887 Congress passed the Hatch Act, which provided federal support for state agricultural research stations, and in 1914 Congress offered aid to the states for developing cooperative extension services.

During the years which followed, the state and national research agencies were under challenge to find political support for research funding, to expand their functions, to find ways to live together, and also to adjust to the transformation from a farm-based economy to an industrial-based agricultural system.

From the beginning there were three types of political supporters for agricultural research: (1) self-designated advocates of the public interest, who wished to improve agriculture through science; (2) scientists, whose interest lay in the creation and funding of research agencies; and

(3) "clients," including farmers, who received some economic benefit from research findings. The public interest advocates and the scientists, usually working together, were instrumental in founding the research institutions; and in the early years their support ensured the institutions' survival. During the twentieth century, the clients or users of research became the major supporters.

Even in the late eighteenth century, a number of public interest advocates had promoted the notion of improving farming through science. The group included editors, scholars, civil servants, and "gentlemen farmers" from landed estates and plantations such as Thomas Jefferson. These people joined in agricultural societies which published journals, held agricultural fairs, and persistently advanced the case for agricultural research at both the state and national levels.[1]

The early scientists who led the effort to establish and develop research agencies tended to be visionaries, propagandists, and political entrepreneurs. Such hybrid spirits nowadays may be a source of embarrassment (or envy) to scientific colleagues, but in those days peer constraints were not so heavy, nor was it unusual for an individual to pursue several careers, perhaps combining the role of scientist with that of farmer, journalist, administrator, or politician.

Three such scientists, John Pitkin Norton of Yale University and two of his successors, developed the first experiment station, from ideas based on observation of prototypes in Scotland and Germany. Norton died at the age of thirty (in 1851), but his mission was taken up by one of his students, Samuel Thompson, and in turn by Thompson's student, Wilbur O. Atwater. Atwater was active in supporting the Hatch Act, finally enacted in 1887, which provided federal support for state experiment stations. He was joined in that advocacy by a number of other interested scientists, including directors of other fledgling experiment stations, who recognized the need to build a body of agricultural science. Meeting at several conferences, they worked out an institutional model that would free scientists from heavy teaching loads and also from servicing the day-to-day needs of local farmers. But in order to gain passage of the Hatch Act, they were obliged to accept a compromise with the presidents of the state agricultural colleges, who insisted that the experiment stations should be adjuncts of their colleges. According to the compromise, the stations would become part of the college, yet would have their own directors and their own national office in Washington.

National research bureaus, as they evolved, also depended on the leadership of one or more dedicated scientists, who were able to create among their coworkers an esprit and a sense of commitment to a selected mission. USDA historians Wayne Rasmussen and Gladys Baker provide a list of exemplary turn-of-the-century scientists who led the way in the development of excellent federal research bureaus:

> With the Department emphasizing the work of individuals rather than offices or projects, still a characteristic of the agency, its formal educational units were established around groups of able scientists. Such brilliant, innovative, and sometimes irascible leaders as Beverly T. Galloway (plant research), Seaman Knapp (extension education), W. J. Spillman (farm management), David G. Fairchild (plant exploration), L.O.Howard (entomology), Harvey W. Wiley (chemistry), Marion Dorset (animal disease), Milton Whitney (soils), Gifford Pinchot (forestry), Victor Olmstead (statistics), Alfred C. True (agricultural education), and W.O.Atwater (nutrition) made the Department the world's outstanding scientific research institution.[2]

These leaders, to gain authority and increased funding, sought support wherever they could find it. Gifford Pinchot, who established the Forest Service, was able to tap the enthusiasm of a conservationist president, Theodore Roosevelt. Scientist-entrepreneurs received support from several secretaries of agriculture, notably Secretary "Tama Jim" Wilson, whose long tenure bridged the centuries (1897-1913). Wilson himself had been a professor of agriculture and director of the Iowa Experiment Station, and also a state and national legislator. Aided by Secretary Wilson's political acumen and commitment to research, the federal research establishment grew so impressively in personnel, functions, and reputation that the state agricultural experiment stations feared being eclipsed by it. In the 1930s, Secretary of Agriculture

Henry A. Wallace provided an environment for rapid growth of national research even as he was creating a "new" department whose major missions went beyond research.

Research leaders could obtain support from powerful congressmen. Louise Stanley, the first head of the USDA's Bureau of Home Economics and Human Nutrition, was able to count upon continuing support from a fellow Missourian, Congressman Howard Cannon, a senior member of the House Appropriations Committee. Powerful legislators like Cannon supported research as a contribution to the public interest. For example, one House Appropriations Committee chairman, James Buchanan, heard Secretary Wallace's call, in 1935, for "much more foundation research to establish laws and principles." Buchanan, who was retiring from Congress, wanted to leave "some little monument to my years in Congress," and therefore he promised Secretary Wallace "all the money he wants for fundamental research, and I will give it to him in lump sum so that he can formulate his own program." From this support, the USDA gained nine new laboratories for fundamental or basic research,[3] and these labs endured as permanent items in the federal budget.

On occasion, leaders of individual bureaus were able to generate public support. A major example was Harvey Wiley, of the Bureau of Chemistry, who came to the department as a chemist in 1883. Wiley invented methods of analysis for research on food adulterants, and because of this research, his division of chemistry became a recognized center of scientific study within the government.[4] As a result of their study, Wiley's group issued a newsworthy bulletin which concluded that food and drug fraud extended to almost every article of food and that the consequences of such fraud fell most heavily on the poor.[5] Wiley, although a federal scientist, became a foremost "public interest advocate." He "dramatized himself as the watchdog of the kitchen, and the incorruptible enemy of the 'whiskey rectifiers' and 'the patent medicine brethren.'"[6] Wiley's flamboyant activities got him into difficulty with Presidents Roosevelt and Taft and with their secretaries of agriculture. But these officers were unwilling to fire him because he had developed strong support within Congress and he had a wide public following. Wiley demanded autonomy with respect to what he should study and what he should publish, as well as in the hiring of subordinates. He fought with other research chiefs, insisting that research within the department should be divided along disciplinary lines and that, therefore, chemical analysis should be conducted by his division only.[7] He resigned, still under fire, in 1912, after department officials decided to impose some coordination upon all the high-flying research chiefs.

The public interest advocate and the scientist-entrepreneur were occasionally combined in the same person, as in Wiley's case. In other cases they were allies in support of research. It should be emphasized that there were many creative scientists who did not gain public reputations but whose work greatly enhanced the reputation of their bureaus.

The third category of supporters — the intended clients or users— was not very visible during the first fifty years of research, from 1860 to 1910. Although the early experiment stations received calls from farmers seeking tests of their soil or water, the scientists who sought public support for research institutions were exasperated by the reluctance of rural legislators to vote that support. Common farmers, themselves unencumbered by book learning, were not able to imagine that agricultural science could help them (and, indeed, throughout the nineteenth century it offered them little help). The common people were more likely to conclude, with their fundamentalist preachers, that science violated God's law. In 1860, however, the Republican party's platform did advocate the development of a Department of Agriculture, as part of a package of agricultural measures. Possibly some Republican party leaders thought voters would be attracted by the research plank, although there is not much evidence, even to the present time, that farm votes can be won by a platform favoring agricultural research.

Even the great farmers' movements of the nineteenth century pursued the promise of social reforms rather than the promise of agricultural science. The Populist movement, when it gained control of the government in Kansas and some other rural states, replaced the top administrators in the Kansas experiment station and agriculture college and increased offerings in economics, in line with Populist views that economic and political reform rather than technological change was the key to a better society. When the Kansas Populists lost power, the Kansas Republicans largely reversed these personnel and curriculum actions.[8]

More typically, politicians intervened in research mainly in pursuit of spoils, seeking to control appointments of research administrators. At the state level, intermittent growth within the agricultural colleges came as a result of efforts by college presidents, experiment station directors, and other administrators to cultivate support among spoils-oriented regents and state legislators.

In the twentieth century, when research institutions were expanding rapidly, two patterns of clientele support aided growth, though neither was conducive to balanced development. One pattern took the form of pressure from producers of a particular commodity. Such special-interest pleading was often encouraged even from within the college by the chairmen of particular academic departments which would benefit from it. Factions competing for research growth thus had wings both inside and outside the colleges. College and station administrators, seeking more generalized support, established "advisory councils" which gave status to farm interests by permitting farmers to participate in research decisions. These farm representatives were in turn expected to help develop a legislative coalition in support of state research funding.[9]

Despite heroic efforts by some college administrations to develop a balanced research program, the major commodity interests in each state commonly insisted that their needs be met first, and they vetoed those kinds of research which might jeopardize their welfare. The pervasive influence of major commodity interests did much to shape, or misshape, teaching curriculums and research agendas as well as the intellectual environment of the colleges.[10]

At the national level also, an interest in a particular commodity was the usual basis of cooperation between bureau chiefs and congressional leaders.

The other pattern of clientele support is associated with the rise of the nation's largest general farm organization, the American Farm Bureau Federation. The Farm Bureau and the extension service grew hand-in-hand. The Smith-Lever Act of 1914 had offered federal matching funds for employment of county-level extension agents who would distribute research findings to progressive farmers. These extension agents were supervised by the state colleges of agriculture. As a means of developing an audience, the agents recruited farmers into county associations called farm bureaus, which in turn affiliated into the statewide and nationwide farm organization. For several decades thereafter, the county agents continued to recruit Farm Bureau memberships, while the American Farm Bureau Federation, for its part, became a supporter of extension and of experiment station research.

The American Farm Bureau Federation, once it entered the national political scene in the early 1920s, secured legislation regulating the activities of middlemen and worked for price-support laws, though on the whole it was quite conservative on social issues. For several decades the Farm Bureau was the most important farm lobby at both national and state levels, and it worked with legislative committees to conciliate and integrate various commodity interests. The research institutions were fortunate in their symbiotic relationship with the powerful Farm Bureau. But it should be stressed that the colleges on their part were also working to integrate agricultural politics, by training leaders and by stressing efficiency and productivity as the means for achieving agricultural well-being. Further, the agricultural colleges were providing an information base and triggering technological change within the economy of agriculture. Extension persons continued to serve as catalysts for political organizations, not only in recruiting Farm Bureau members, but also in serving as secretaries for the developing commodity organizations.

Research scientists became enmeshed in this informal network of power, even though research organization was formally decentralized and independent. Commodity interests worked closely with researchers in the colleges and federal bureaus. The agricultural subsystem, through congressional appropriations and through the Farm Bureau linkage with colleges, selected the building blocks for research—the institutional leadership, the research facilities, the mix of scientific personnel, and, indeed, the training programs. The subsystem also imposed the rule that research should be useful to commercial agriculture, and in no way embarrassing to or competitive with it.

CONTROLLING RESEARCH THROUGH
LAW & ADMINISTRATION

Intermittently there were efforts to improve the structure for managing research. Sometimes the intent was to find a structural shield against the farm pressure groups and other informal influences. The national office of experiment stations, for example, was often effective, as its creators had hoped, in protecting the stations from "the vicissitudes of state politics".[11]

Formal coordination was also urged as a way of resolving conflict and avoiding duplication of effort, as well as fostering cooperation between agencies—particularly between the states—and assuring that research is well managed and directed toward public goals.

Recommendations for better coordination and management have come in chorus from respected and knowledgeable evaluating agencies, including the General Accounting Office, the House of Representatives' Science and Technology Committee, the congressional Office of Technology Assessment, the president's Office of Management and Budget, committees of the National Academy of Science, and even from a USDA review committee composed mainly of agricultural research administrators.[12] Countering these recommendations, the decentralized research agencies argued that they were already implementing better management and that there was a danger in overmanaging research; that efforts to state precise objectives and to measure achievements are a waste of scientists' time; that scientists are unlikely to duplicate work unnecessarily because there is no reward for doing so; that scientists chart their own objectives as they uncover gaps in knowledge; and that scientists best serve the public interest by following their own intuition.

Behind the arguments over the virtues of coordinating management were struggles over who should control agricultural research. After 1972 there were several new legislative directives, new machinery for planning and coordination, agency reorganizations, and new legislative mandates. Virtually all of these were perceived by the congressional appropriations committees, by the state experiment stations, and by agribusiness groups as efforts to centralize power in the national executive branch.

Figures 2.2 through 2.5 show the various efforts since 1860 to construct a centralizing hierarchy for agricultural research institutions. Looking first at the hierarchy for state institutions (fig. 2.2), we see that the Hatch Act of 1887 created an Office of Experiment Stations (OES) within the USDA. Its first director was Wilbur Atwater, the last of the three Connecticut professors who championed the experiment station idea. The OES and its successor agencies functioned as national advocates for the state stations, especially in their competition with federal research bureaus for federal funding. In addition, the OES made a seminal contribution to the management of research by overseeing the development, by 1938, of "contracts" for all research projects. This system, later refined under the name of the Current Research Information System (CRIS), made it possible to aggregate and analyse on-going federal and state agricultural research.

The OES has also provided consultative services and liaison for the experiment stations.[13] But there was relatively little effort at any time to guide or evaluate experiment station research, despite the fact that federal funds were being used. Charles Hardin, who examined the political environment of experiment station research during the 1950s, reported that the experiment stations experienced little federal control.

In the years before World War I, it had become a principle that the stations were to be independent of federal direction. Indeed, a com-

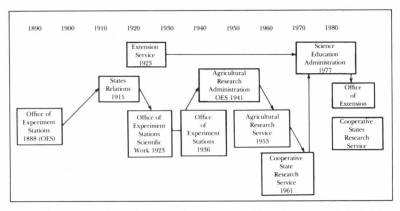

Figures 2.2.: National Offices of State Experiment Stations and Extension. (*Source*: Gladys L. Baker, Wayne D. Rasmussen, Vivian Wiser, and Jane M. Porter, *Century of Service: The First One Hundred Years of the United States Department of Agriculture* [Washington, D.C.: Centennial Committee, U.S. Department of Agriculture, Government Printing Office, 1963]).

mittee of experiment station directors was to be consulted before any federal actions were taken which might affect the stations. The department's mode of supportive involvement in experiment station research was called "participation without control."[14]

Federal legislation authorizing research funds contained few stipulations as to what research should be done. The founding law—the 1887 Hatch Act—said experiment stations should aid in "acquiring and diffusing among the people of the United States useful and practical information on subjects connected with agriculture" and should help "to promote scientific investigation and experiments respecting the principles and applications of agricultural science." Funding for experiment stations was occasionally earmarked—for example, for "original research" in the Adams Act of 1906, and for "marketing research" in the Research and Marketing Act of 1946.[15]

Of course the Office of Experiment Stations did require that federally funded research projects be specified and described, and the OES sometimes felt obliged to protest against the improper use of research funds for teaching or purposes other than research. In the nineteenth

24

and early twentieth centuries the OES tried, with little success, to emphasize basic research and to prevent the use of national funds to develop state substations. Again in 1977-79 the USDA undertook to impose priorities in the use of Hatch Act funds; the resulting outcry from experiment stations made it clear that this initiative, though authorized in 1977 legislation, was at odds with previous experience and current preference.

Therefore, the major questions resolved by federal law and formal agreement have been (1) how to divide funds and functions between federal and state research establishments, and (2) how to divide federal research funds among the states. Between state and federal establishments the proclaimed relationship, and the dominant mode, is that of a partnership, "just like Mr. Sears and Mr. Roebuck."[16] Rivalry occasionally breaks forth, and at times the two act like "the Gingham Dog and the Calico Cat", ready to eat each other up. When federal agencies flowered in the nineteenth century, the states, fearful of becoming "mere outposts to a monolithic federal research establishment,"[17] argued before a congressional committee that the USDA's growth should be checked; Secretary of Agriculture Tama Jim Wilson, himself a former state experiment station director, conciliated the directors by supporting increased experiment station funding, even as he continued to develop the federal research agencies.

In 1906 the experiment stations formed an organization to speak for their interests, the Experiment Station Committee on Organization and Policy (ESCOP). A group was formed for cooperative extension also, the Extension Committee on Organization and Policy (ECOP), and both affiliated under the National Association of State Universities and Land Grant Colleges (NASULGC). Recently, when both state and federal agencies were threatened with reduced funding, state leaders hired a lobbyist to argue their interest. There are advocates from each level—more often from the states—who argue that their own level is most productive and therefore should receive relatively more—if not all—federal research funding.

With respect to the division of federal funds among the states, the

Hatch Act formula of 1887 provided a flat $15,000 per state. Funds were first distributed on the basis of total state population, and subsequently on the basis of "rural" population. Supplemental funds have since been added to those granted under the rural population formula, including funds for the "negro" colleges that were instituted in most Southern states in the 1890s, and also funds for grants awarded under a competitive or merit selection process. The competitive grants reflect an acknowledgment of the fact that the relative merit of state institutions is not correlated with the relative size of the state's rural population.

The federal research structure has presented similar challenges in management and administration. As developed by the scientist entrepreneurs, it initially took the form of a number of functional agencies or bureaus which became full-fledged bureaus, or statutory bureaucracies, after the turn of the century (fig. 2.3). Although most of these were ultimately merged in the Agricultural Research Service in 1953 and then in the Science and Education Administration in 1977 (abolished in 1981), at least two of the bureaus continued to have independent status. One, the Forest Service, has enjoyed such a good reputation that it has maintained nearly complete autonomy within the USDA and has successfully resisted several major efforts to transfer it elsewhere. Another agency, the Bureau of Agricultural Economics, was abolished in 1953, despite its formerly distinguished reputation, because conservative legislators considered it to be an advocate of social change. An agency for economic research (the Economic Research Service) was recreated within the USDA in 1961, and in that same year the Statistical Reporting Service (SRS) was formed in an effort to improve crop estimates. As figure 2.4 shows, ERS and SRS were merged in 1977 into the Economics and Statistics Service and then separated again when ESS was abolished in 1981.

INTEGRATING RESEARCH AGENCIES

As in a recurring nightmare, every effort thus far to integrate public

26

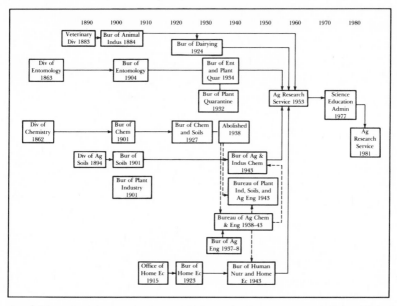

FIGURE 2.3: Organization of Federal Agencies in Agricultural Research. (*Source*: Gladys L. Baker, Wayne D. Rasmussen, Vivian Wiser, and Jane M. Porter, *Century of Service: The First One Hundred Years of the United States Department of Agriculture* [Washington, D.C.: Centennial Committee, U.S. Department of Agriculture, Government Printing Office, 1963]).

agricultural research institutions has proven unsuccessful (see fig. 2.5).

In 1921 Secretary of Agriculture Henry C. Wallace established a "director of scientific work," whose task was to "bring about more complete cooperation in scientific research work in the Department and the various state experiment stations and colleges."[18] The Office of Experiment Stations was downgraded, temporarily, to a subbureau within the secretary's office.

The science director's position was abolished in 1934, an admitted failure. In 1936 Secretary Henry A. Wallace (the son of former Secretary Henry C. Wallace) designated the chief of the Office of Experi-

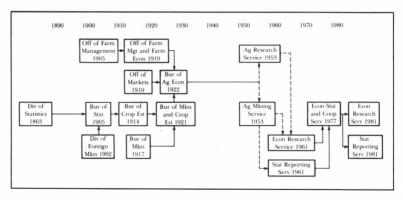

FIGURE 2.4: Organization of Federal Agencies in Economic Research. (*Source*: Gladys L. Baker, Wayne D. Rasmussen, Vivian Wiser, and Jane M. Porter, *Century of Service: The First One Hundred Years of the United States Department of Agriculture* [Washington, D.C.: Centennial Committee, U.S. Department of Agriculture, Government Printing Office, 1963]).

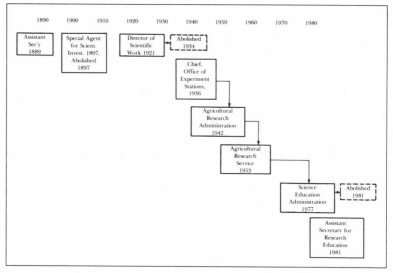

FIGURE 2.5: Overall Coordination of Agricultural Science by the U.S. Department of Agriculture. (*Source*: Gladys L. Baker, Wayne D. Rasmussen, Vivian Wiser, and Jane M. Porter, *Century of Service: The First One Hundred Years of the United States Department of Agriculture* [Washington, D.C.: Centennial Committee, U.S. Department of Agriculture, Government Printing Office, 1963]).

ment Stations as "director of research" for the federal bureaus as well. This effort to achieve coordination also failed.

In 1942 Secretary of Agriculture Claude Wickard created a "holding company" for federal research bureaus and the state experiment stations in the form of an overhead administration which he named the Agricultural Research Administration. This superagency appeared effective enough to be singled out, in a 1947 report by the president's Scientific Advisory Board, as "the outstanding illustration of a single departmental unit charged with nearly all the scientific research and development work of the department."[19]

The Agricultural Research Administration was replaced in 1953 by a new structure, the Agricultural Research Service (ARS), which embraced only the federal research bureaus and did in fact bring these research bureaus under a central administration (the Office of Experiment Stations was reincarnated as a companion agency, ultimately under the name of the Cooperative States Research Service). One observer described the merging of federal research bureaus into ARS: "It was quite a spectacle, and a traumatic experience. Those bureau chiefs who did not resign or commit suicide had a terrible time fitting into the new situation."[20] Although the 1953 reorganization abolished the bureaus, a shadow of their former status was maintained in the form of line officers called "investigation leaders." A more thoroughgoing reorganization occurred in 1972, under which the line officers were given geographic rather than subject-matter jurisdictions. Subject areas were thereafter represented only within the structure of a "national program staff" which served in an advisory relationship to the area and regional line officers. A major purpose of the 1972 reorganization, in accord with Nixon administration policy, was to permit decisions to be made at the regional and area levels.

This more complex administrative network provided opportunities for coordination and for leadership on the part of both line and staff officers. Under the 1972 structure, the secretary of agriculture also became more important in framing research budgets, a role which, under the former system, had fallen to bureau chiefs, who had dealt directly with senior rural legislators.

Many advantages were presumed for the new system. It would permit a multidisciplinary approach to problems; make researchers accountable to the secretary and the president; and protect the agency's fundamental research commitment against harassment from Congress and local groups. Also, the fruits of its research could be distributed among all citizens, not just among farmer and agribusiness clients. Thus the federal research agency's resources would, for example, be more accessible to federal "action" agencies such as those which protect food safety and those which educate citizens about nutrition.

Besides integrating federal research agencies, the ARS, as reorganized in 1953, was given responsibility for coordinating the state experiment stations (note fig. 2.2), but by 1961 the experiment stations again had their own separate agency—the Cooperative States Research Service (CSRS).

As a step in improving coordination between state and national research and extension, in 1971 the USDA and the National Association of State Universities and Land Grant Colleges developed a "memorandum of understanding," under which a joint planning and coordinating committee, called the Agricultural Research Policy Advisory Committee (ARPAC), was established.[21] ARPAC had two chairpersons—the assistant secretary of agriculture and a state leader designated by the colleges—plus eight additional representatives from the state system and eight from the federal system.

In 1977 a superagency was created for agricultural research—the Science and Education Administration (SEA). Under SEA were combined the ARS, the National Agricultural Library, the Cooperative States Research Service, and the Office of Extension; and SEA was also given the additional mission of improving the training of scientists. SEA's director was provided with staffs for the coordination and evaluation of research. In addition, a joint council of the research-producing agencies was created (superseding ARPAC), which included representation from organizations outside the USDA and the land-grant institutions. Further, an advisory committee of "users" was established, with a membership composed of environmentalists, agribusiness representatives, farmers, and food safety and nutrition advocates. Both groups were placed within the research decision-making structure. In addition, the 1977 law called for an interdepartmental Subcommittee on Food and Renewable Resources which would review and plan research in the light of government-wide objectives.

The new structure was expected to remedy a number of shortcomings cited by various evaluating groups, including inadequate cooperation between state and federal administrations, uneven quality of research, and reluctance to address such nonagricultural research problems as human nutrition and consumer safety. It also gave the secretary of agriculture more tools for influencing research policy, even as it established a rich competition among the factions represented in the decision-making processes. Informal controls from Congress and from the industry continued, however, and former bureaucratic structures remained active beneath the surface. Within the planning staffs and on the joint council, state and federal representatives refused to lay their own plans out on the table. Thus it was not surprising that the priorities which were finally established in the 1979 executive budget were almost totally reversed by the congressional appropriations committees.

The newly integrated system had produced neither a cooperative spirit nor a sense of exciting new missions. Rather, agricultural scientists perceived their institutions to be—temporarily at least—in a state of confusion and decline. In 1981, the Science and Education Administration was abolished, and the components regained agency status under their former names.

CHAPTER THREE

Agricultural Scientists at Work

Despite problems of funding and organization, agricultural research in this country has yielded amazing fruit. Whether in federal research facilities or on college campuses, in the field or in the laboratory, agricultural scientists continue to seek answers to the problems that beset agricultural production as well as those related to health and safety. In effect, they constitute a distinctive subculture within the larger society, a subculture which is largely defined by reference to where its members work, how they work, and how they are trained, and which we will examine in this chapter.

FEDERAL RESEARCH INSTITUTIONS

The showcase of American agricultural research is at Beltsville, Maryland, just outside of Washington, D.C. At the sprawling Beltsville complex, foreign and domestic visitors circulate along narrow roads and stroll through quaint buildings to observe milk cows walking spraddle-legged around their bulging udders, tall windows of "compost" refined from Washington's daily output of eighteen hundred wet tons of sewage sludge, wilted tomato plants which have been intentionally exposed to big-city air pollution, and the world's most complete "collection" of fungi, insects, seeds (for example, fourteen thousand varieties of rice), and nematodes (tiny soil worms). Beltsville, the work-place of about one-fifth of all federal agricultural research scientists, is organized into sixty-two laboratories, each usually staffed by subgroups of fewer than ten persons. Beltsville is "one of the largest and most diversified research farms in the world,"[1] although its long-eroded soils are among the least productive in the country.

Beltsville became a center during the Great Depression, when several Washington-area research activities were aggregated there. Many scientists think Beltsville was not a good idea—partly because it violated the principle of "pork barrel" politics, under which research funding is obtained by locating laboratories in the districts of powerful legislators—and in recent times there had been thoughts of disbanding it. Its unimproved, pre-World War II physical structure is strikingly Spartan alongside the impressive quarters of other Washington bureaucracies. But it seems now to have been decided that Beltsville will remain open, in part because a great agricultural nation needs a showcase, but also because Beltsville is a tremendously productive place.

Beltsville is one kind of workplace—a large one. At the other end of the size spectrum, in the high-plains small town of Akron, Colorado, is a federal research laboratory with a "multidisciplinary" team, consisting of an agricultural engineer and three soil scientists, whose mission is to study the soil and water problems of the area.

There are dozens of tiny installations like that at Akron, partly in response to the many unique agricultural environments in this country. (Indeed, the Akron laboratory helps compensate for the earlier failure of the Great Plains states to place state experiment stations out on the high plains, a failure which delayed the development of the Great Plains grain and livestock economy.[2]) But it was politics as much as unique geography which caused the proliferation of federal research sites. Farm groups insisted on locations that would guarantee attention to their particular commodity, while powerful senators were continually sponsoring new installations for their states. In 1977, for example, Congress created a new laboratory in the Senate majority leader's remote West Virginia hometown.

There are now more than 140 federal research "locations," of which a few, including Beltsville, are large or medium-sized research centers. About fifty are joined with land-grant colleges, and most others have fewer than ten scientists, despite the judgment by research administrators that scientists in these small labs suffer from a lack of the technical support, inspiration, and flexibility found in larger installations. In recent years, federal research administrators have closed out a net of 25 of the small labs as their congressional sponsors retired from office. Also, the U.S. Forest Service has successfully moved its scientists from small remote forest labs into regional centers or onto campuses.

Several "medium-sized" federal research centers are dedicated to specific research missions. An example is the Human Nutrition Research Laboratory (with 26 employees) at Grand Forks, North Dakota. Another is the Richard B. Russell Research Center (with 168 employees) at Athens, Georgia, which serves the research needs of the southeastern states. The Russell institute exemplifies the results that can be achieved through a judicious balancing of interests and resources.

HARRY NEUFELD AND THE
RICHARD B. RUSSELL AGRICULTURAL RESEARCH CENTER

Until he retired recently, Harry Neufeld, who speaks with the softness of a man recovered from open-heart surgery, was in charge of a marble-inlaid research palace named for Sen. Richard B. Russell, the powerful "dean" of the Senate for many years. Under Neufeld, The Richard B. Russell Agricultural Center became a successful regional institute, though not without difficulty. Senator Russell died the year after he had secured funds for a beautiful research facility in his home state, and in his absence Congress failed to appropriate funds to staff it. The nearly empty building was an embarrassment to the citizens of Athens, Georgia, who lobbied for three years to get it into operation.

The center and two nearby federal research installations, also administered under the center director, work with state scientists at the University of Georgia (also located in Athens) and with "cooperating" farmers throughout the Southeast region. Most of the cooperating farmers are big ones who "lend" land and other resources for multi-year research projects.

The center has improved the processing of Georgia cling peaches. It has developed a new process for Georgia's "steamed" pecans. It has worked out a new way of harvesting water chestnuts, thus restoring that domestic product to our grocery shelves. It is rapidly mechanizing the broiler processing industry, a conversion which will reduce the amount of water used, reduce the amount of pollution from wastes, and further reduce the risks of salmonella (a disease which consumers can contract from improperly processed foods). Because the mechanization of broiler processing will put large numbers of Georgia women out of work, the Economics and Statistics Service is doing a technology assessment study to compare this "economic cost" with the "economic benefits."

Dr. Neufeld's researchers worked with a private company to develop the now famous Gold-kist fireplace log made from peanut hulls. They are studying the composition of sunflower seeds to gain greater utilization, and they are locating the carcinogens in tobacco, hoping to produce a safer cigarette. Some of the center's scientists are studying certain wild plants within the region which are extraordinarily efficient in mining energy from the sun. Their hope is to breed this efficiency into such plants as rice and soybeans.

Harry Neufeld thinks that the southeastern United States may replace the West as the nation's great vegetable basket because it has an abundance of water. He has a dream that the red soils of Georgia will someday produce three crops per year, and although he is no longer on active service, he continues to love his scientists and his center, and he cherishes the memory of Sen. Richard B. Russell. It would appear that if the center can continue to produce the results it achieved under his stewardship, his dream might well become reality.

UNIVERSITY EXPERIMENT STATIONS

Most agricultural researchers are located on or near college campuses.

The directory of professional workers at the state agricultural experiment stations and other cooperating state institutions lists twenty-six thousand persons, well over half of whom are Ph.D.'s in some branch of agricultural science.[3]

Most agricultural scientists at the state land-grant colleges are also part-time teachers. For example, the work time of a young assistant professor at the University of California–Davis may be allocated 71 percent to research, 29 percent to teaching. He or she may be responsible for teaching one undergraduate class, with encouragement to develop one new upper-division course.[4]

Comparisons are made, favorably and unfavorably, between the part-time state researcher and the full-time federal scientist. One federal scientist argues that the experiment stations' part-time researchers do not have time to do serious projects, and that their work escapes evaluation:

At the university you don't have to go far to see that the students come first, so you always are putting off things, having to treat the person first and the research second, and some professors never get around to doing any research even though they are paid half to three-fourths for research; and the research that they do may not go much beyond the superficial.

Another disadvantage of universities is that there may be more chances for unproductive researchers to stay; because they can claim that they are teaching even though if you examine their teaching, the students will say that they are involved too much in research. Another "out" is that they have extension activities that are superimposed on their other jobs.

It is also argued that since the universities are organized by academic disciplines rather than according to research problems, researchers find it difficult to form multi-disciplinary teams that can address major problems.

In favor of the experiment stations, it may be noted that young scholars are on probation for five or six years, after which only the devoted and capable researchers are likely to be kept.[5] Further, the universities are able to save money by using graduate students as technicians while they are being trained as scientists. The latter point, however, does not really apply to the less developed universities—Arkansas, Delaware, Georgia Tech, and Maine, for example—which had produced none of the scientists employed in federal research as of 1977. Moreover, a number of other land-grant colleges had each produced fewer than ten federal Ph.D. scientists at that time. In contrast, five universities—California, Iowa State, Minnesota, Cornell, and Wisconsin—had produced a total of 637—more than one-third—of all the federal service Ph.D. scientists.

The fifty-five state experiment stations vary in size and reputation, with some of the smaller and less distinguished institutions receiving relatively large federal grants because their states have large rural populations. Even special research funds, such as those for corn leaf blight, are widely distributed so as to include small institutions. As one observer of leaf-blight funding noted: "Funds went to some states because legislators from those states sat on the Agriculture Committee. Mississippi was selected because of political pressure—Representative Whitten. We go along with Mississippi, or no money for anyone."[6] Such political considerations tend to skew research priorities and to make efforts to achieve coherence in budgeting and planning even more difficult.

THE PROCESS OF RESEARCH

During the months of June and July in 1977, federal research administrators were agonizing over which of their three thousand projects must be put on the "hit list," probably to be killed under President Carter's new "zero-based" budget approach. Meanwhile, the research agency's news service was describing some new research projects, and some new findings from existing ones. Bee researchers from federal research agencies and from Ohio State University were undertaking a one-year study to find out why some bee colonies disappeared during the winter while others did not. Other federal bee scientists were in Brazil trying to stop the migration of "killer" bees toward the U.S. border. Among the reported successes of other projects were the findings that small farmers in Mississippi could make a good living raising rabbiteye blueberries; that dogs could be trained to sniff out cows "in heat" and ready for artificial inseminiation; that "educated" lymphocites could recognize and attack disease organisms—a possible breakthrough in basic knowledge; and that a single strand of "electric" fence was effective in keeping coyotes away from sheep.

Starts, stops, and successes make research seem like a track meet. Indeed, the spectacular successes in research often come from intense competition among scientists who are chasing the same hound in similar laboratories, a phenomenon which budget-cutters may see as wasteful. But in the "compete and communicate" style of public research,[7] these laboratories replicate and leapfrog one another in their rush to success.

Along the road of scientific discovery a few milestones may be placed. One is the Nobel Prize, given to Norman Borlaug for his successful adaptation of short-stemmed wheat. Individual achievements such as Borlaug's, however, are usually capstones in an edifice to which dozens of scientists have contributed, and which may still be tentative and incomplete. Borlaug's achievement, in fact, was less one of discovery than of building a research institution which could subsequently pursue development of the "miracle wheat."

Plant geneticists, in particular, are likely to feel that the research process has no beginning or end—no starts, stops, and wins—but is rather a continuous process—a circle, or winding staircase. Therefore, the usual policy controls, which are bound in time frames defined in terms of priority planning, one-year budgets, five-year objectives, and evaluation of outputs, are likely to prove awkward for them. Plant geneticist Charles Lewis points out: "We have a problem with that person who says, 'You've got to write a work unit of five years' duration,' because if you are talking about plant breeding, it is not a five-year period but a continuum."

Lewis explained that there are several tiers of activity in producing a seed with high capability. The base activity is to build a bank of all known seeds. Lewis has helped to create seed banks for the United

States, including an agency to exchange seeds with other countries, a "sealed" bank for long-term storage, four regional storage and use centers, and a number of state and private centers for major crops.

At the next level in plant breeding is the "screening" process, the purpose of which is to identify plants having desirable characteristics and to determine which pairings of plants will accept cross breeding.

Then comes the breeding itself for a particular goal (for example, to incorporate plant resistance to insects). The task of breeders is to continue all the desirable traits within a single stock—as Lewis put it, "to get all the coons up the same tree."

"Finally, we have genetic fine-tuning—a constant readjustment of what we plant in the world to what is demanded of us in order to achieve certain things, including a growth in productivity, of which plant breeding accounts for about a 1 percent increase per year since the turn of the century."

As Lewis sees it, "The problem is, as in the past, that nobody can see that you're trying to build this cathedral."[8]

So scientists are likely to stress the continuity of the scientific process, indeed to say that it is more than a process. It is the development of institutions, the building of a cathedral.

In the game of science there is not orderly development but frenetic competition—for results, and also for funds, which must be obtained from sponsors who are not much interested in cathedrals. These include agribusiness clients who want to avoid being inconvenienced by a pest or disease, a new government regulation, a competitor, or the farm workers' demand for union representation. When an inconvenience becomes a crisis in the clients' perception, they may release a river of research funds with the promise of career rewards to those scientists who fulfill their expectations. Agricultural researchers, in the process of satisfying the need, seize the opportunity to add a wing to their jerry-built cathedral. The way the research process works in such circumstances is illustrated in the story of Graham Purchase and the sick chickens.

GRAHAM PURCHASE, ET AL., & THE SICK CHICKENS

This is a story about a typical chicken: a sick chicken. It is the story of two laboratories—Houghton, England, and East Lansing of Michigan—that did something about the sickness. It is also about a scientist who took a leading role.

An egg for breakfast and chicken at any other meal appeal to our tastes, and even to nutritionists' notions of what is best for most of us to eat. For many years the chickens that produce these foods have been likely to suffer from viruses which cause malignant tumors in the lymph glands. Some chickens die from these viruses, and the others are debilitated. So poultry scientists have been keen to find remedies for "big liver disease," "gray eye," and "paralysis," as these viruses were once named because of their symptoms.

In 1937, representatives of the poultry industry met to deal with this problem. At the meeting were distinguished poultry scientists, chick breeders, poultry organizations, poultry journal editors—everyone was represented except the farm wives, who were at that time the backbone of the chicken and egg industry. These leaders asked Congress to establish a poultry research laboratory, which was created in 1939 at East Lansing, Michigan. Meanwhile, in England, private groups were evolving a similar chicken research institute at Houghton, England, which was ultimately funded by the British government. Both Houghton and East Lansing devoted their research to the chicken viruses, seeking mainly to breed chickens highly resistant to these diseases.

During the 1950s, millions of backyard chicken houses were replaced by large, confined, "intensive" operations controlled "vertically" by producers of chicken feed or by supermarkets.[9] The new, large-scale producers were politically astute, and they were more sensitive to disease losses. Their concern about disease loss was heightened with passage of the Poultry Inspection Act of 1961, under which federal inspectors could condemn dressed broilers which bore obvious marks of disease. Within the industry there was great agitation for a solution. Legislators from areas where the chicken industry had concentrated, particularly in the southeastern United States, demanded a federal research program to find a remedy for the disease. The main research decision-makers were the Congress's agricultural appropriations sub-committees, which moved to establish new poultry laboratories at Athens, Georgia, and at Mississippi State College. Congress also authorized a large expansion of facilities at East Lansing, and it lavishly funded scientists at these and other poultry laboratories during the following ten years. The funds created a sense of urgency and stimulated a race between the front-running institutions—Houghton and East Lansing.

The principal remedy turned out to be a vaccine against the most devastating virus, Marek's disease. The vaccine was developed by an East Lansing team under the leadership of Graham Purchase.

Purchase himself was British, born in Rhodesia, raised in Kenya, and educated in veterinary medicine in South Africa, following his father's profession. After coming to the United States, he served at East Lansing for three years, then went to Houghton for a year to tool up for further study of Marek's disease.

Five years later, in 1971, Graham Purchase and his colleagues at East Lansing announced an effective vaccine for Marek's disease, a vaccine which is now administered in the United States to virtually all chickens on the day they are hatched. Although the problem of Marek's disease has not been eliminated, since the introduction of the vaccine the number of broilers "condemned" by poultry inspectors has decreased from 1.6 percent to less than .2 percent,[10] and egg production per hen has increased by 10 percent.[11] Profits to the manufacturers of the vaccine, in 1974 alone, were $3.7 million.[12] In the first four years of use, the computed savings to the poultry industry were $615 million, compared with a research cost of only $31 million for all U.S. public agencies during all previous years. As a result of this achievement, Purchase and his colleagues received many awards, and Purchase was promoted to an administrative position in Washington, D.C.

The chicken industry, which had so desired this solution, went into an immediate slump resulting from overproduction. In fact, Purchase, in describing the satisfactions of success—financial rewards, promotion, and the excitement of having admiring visitors in his lab—also mentioned that "when the price of eggs nine months later was low, I could tell my wife and friends that it was directly a result of overproduction of eggs because we had done such a good job."[13] Purchase regretted the fact that low prices had forced many producers, most of them farm wives, to board up their chicken houses.

There were many contributors to the Marek's disease vaccine; indeed, there was no real beginning in this discovery process. The cause of Marek's disease had been identified in 1962 as a new (separate) disease. Houghton scientists had made a major discovery in 1967—that the tumors were caused by a virus. The East Lansing scientists made the same finding independently, one month later. Both East Lansing and Cornell University scientists soon discovered that this virus was passed along like smallpox, from surface sores in the chicken hair follicles.

Houghton then developed the first vaccine, made of weakened Marek's disease virus and now commercially used in a few countries. Purchase's more practical vaccine was developed from a mild turkey virus. The turkey virus had been discovered by a Wisconsin team who apparently did not recognize that they had, in effect, found the basis for a vaccine. Throughout the fast-paced competition to find a preventative for Marek's disease, both the Houghton and East Lansing labs maintained a total openness with others about their experiments. It was this openness that enabled East Lansing to be "first" with the turkey virus vaccine. Purchase explained: "We gave out turkey virus to anyone who asked for it. If we had withheld it, then the Wisconsin people who discovered it would have known to exploit it, you see, and the virus now used worldwide would have been theirs, not ours."

Purchase's main contribution was as group leader of ten scientists. He organized a follow-through from ideas generated in group discussion. "We had a large number of experiments, one after the other, and we had to make sure we were at the right places at the right times." His thoroughness paid off in securing the government's quick approval for the use of the vaccine, since the group, even as it developed the vaccine, had run all the necessary tests for unanticipated side effects from the vaccine.

Work goes on, some of it generated from the findings: human cancer researchers proceeded with new understanding that viruses can cause cancer (although they could not easily emulate its solution by vaccinating humans at birth for diseases which strike at middle age). The East Lansing lab continues to work on the leukoses, and on Marke's disease, too, because the vaccine is not totally effective. Agricultural scientists typically continue to play cat-and-mouse with a disease or pest that they have partially, and perhaps only temporarily, controlled, in hope of complete success.

THE TRAINING OF AGRICULTURAL SCIENTISTS

Having looked at where they work and how they work, let us turn now to the scientists themselves—or more particularly, to the process which prepares them for their work. By some accounts, the training and experience of agricultural scientists has been quite narrow. "Our scientists are politically narrow; they do not have perspective," observes a prominent agricultural experiment station director. He, like many other agricultural educators, has repeatedly urged a broadening of college curriculums.

Until recently, most of the students in the colleges of agriculture (which produced most of our agricultural scientists) had been raised on farms, and according to a recent survey, a large majority of agricultural scientists in most fields had farm backgrounds.[14] Currently practicing agricultural scientists grew up during the period 1910 to 1950, when there were relatively few "big" farms, but much difference among farmers. Rural children thrust early into hard labor, they disliked pitching hay, milking cows by hand, scooping grain, and dawn-to-dusk tractor driving. They developed no sympathy, either, for the lazy farmers who went fishing when they should have been cultivating, or for the farmers who did things "the hard way," resisting the use of new machines and practices. They knew it was not hard work per se that provided the family with income and status; rather, it was the expeditious use of their labor, their land, and all other resources.

Some of these young rural people chose to seek status and income by some means other than farming and, for various reasons, chose agricultural science. Plant scientist Charles Lewis, for example, grew up on a small, nearly self-sufficient cotton and general farm, knowing that "there wasn't room for me to make a life on that farm." Nevertheless, he went into an agriculture curriculum at college, because he simply did not know about other occupations. "When you've grown up in the country or in a small town, your perspective is not wide as to what the world has to offer. In our town we had an old doctor, and we knew about teachers and preachers, but I don't think we even had a lawyer. There is a host of occupations that don't show up in a place like that."

Ray Webb, who now develops potato varieties important to East Coast farmers, became a scientist because he was physically too small to do the heavy work demanded on his family's farm, and also because he was drawn to the mystery of instrumentation as used by agricultural engineers laying terraces in his area. He was much impressed by the U.S. government "logo" painted on the doors of their vehicles. Once in college, Webb was happy to earn his way with hundred-hour work weeks in the vegetable plots of his professor, who passed along to Webb his skills, his aggressiveness, and his ambition to become the "Luther Burbank of the South."

Those boys and girls who became scientists or home economists received their formative educational experiences at the state land-grant schools, where they earned degrees in one of the several academic disciplines offered by the colleges of agriculture. Nine of every ten Ph.D.'s in the USDA's Agricultural Research Service, as of 1969, had received that degree from a land-grant school.[15] As Don Price notes, "the land grant colleges and the associations of various kinds of agricultural scientists maintained an important influence on the Department of Agriculture, supplied most of its career personnel, and generally provided the intellectual leadership for national agricultural policy."[16] These colleges also produced the farmers and businessmen who returned to become leaders of the rural community and to organize the agribusiness firms.

The land-grant colleges were single-minded, revolutionary institutions. At first they refused to teach courses in the airy fields of literature and history, or even in theoretical science. Such useless courses were the fare of elitist private colleges. The "aggies" at the "cow college" created their own academic disciplines, such as agronomy and dairy science. In cases where the land-grant college became the state university, the college of agriculture had its own campus, or in any case, led its own somewhat isolated life on the "back forty." Because farming and associated industries were generally considered to be low-status occupations, agricultural people insulated themselves in a subsystem which developed its own status and achievement symbols.[17]

However, there were always scholars in agriculture working to create a broader perspective both in teaching and research. At Iowa State College, for example, the theoretical sciences were smuggled in. A distinguished chemist at Iowa State, Henry Gilman, recalled that "there was no such thing as 'physics,' only 'applied physics,' " and since only "relevant" research could be done, Gilman's master's thesis was entitled "The Utilization of Agricultural Byproducts." The title, he explained, "was something to placate someone whom we don't know." A subtitle revealed his real subject—a fundamental research problem dealing with the behavior of a particular chemical.[18]

Strict constraints on learning and research were breached by Gilman and others, through reading, travel, and through conversations with visiting scholars, but these constraints set the tone for the campus as a whole. Undergraduate students were immersed in practical courses in "animal husbandry" within a field such as "farm operations," and they had their social contacts through the "farm op" club.

At graduation, they were separated into occupations—some going into agricultural businesses, some going back to the farm, and some proceeding on to graduate school. On leaving graduate school the scientists were again distributed—some to the universities, some to the federal research agencies, some to private industry. But the colleges had already established their graduates' reference groups, which were maintained following graduation through conferences, correspondence, journals and trade magazines, and exchanges of visits and favors. Indeed, these college reference groups were virtually the only professional contacts some scientists had. One critic of the research establishment, Susan DeMarco, tried to find a conspiracy of special interests among researchers, college administrators, and the principal agribusiness firms; she finally concluded that the establishment was less a conspiracy than a fraternity of "good old boys."[19]

Agricultural scientists are trained for narrow specialties within particular disciplines (which have themselves been shaped in part by previous user demand); but the scientists, as they mature, are expected to develop a greater breadth of interest. Young scientists, while still in graduate school, find themselves sliding down a greased skid called "specialization," which will place them within a research group whose mission has been previously established and in which they will play a well-delineated role. For example, a graduate student in economics who is attracted to the work of a professor whose competence is in judging the value of agricultural land may, on receiving the Ph.D., be employed by the USDA on a research team whose function is to provide current information to land sellers and purchasers on the present and projected values of farm land of different types.

In a sense, this first position constitutes a "box canyon" which the young scientist will be allowed to explore. But in time creative scientists are expected to grow beyond the role for which they were hired. James Kendrick, experiment station director at California-Berkeley, has described the process of growth:

> This is what we do: we find an area that needs work; we create a job with describable competence; then we seek a person and turn him loose and he will ramble around within that area. For example, my first job was to study vegetable diseases—and root rot was the big one, so I studied it. I began to branch out and concentrate on soil-borne fungi. I was still doing my job at bottom. I think this is rather ideal. Remember that each of these things to

46

which he is assigned will not be a problem after forty years or so, so he should branch out and expand his interests. Anybody who has to come and ask the boss what to do next is not imaginative enough to be employed here.[20]

The bond among agricultural scientists is rather distinctive in being reinforced by common heritage, common academic subdisciplines, linked institutions, and dependence upon the same political supporters and funding sources. These bonds support a "family" environment in which achievement is defined, competition is keen, and successes as well as failures are highly visible. There are "father figures" who support participants even as they judge them sternly. One such father figure is soil physicist Bill Raney, recently retired, a genial Mississippian whose official role within the USDA's Agricultural Research Service was to coordinate soil, water, and atmospheric research. Informally, he also served as counselor for many soil physicists. He published a newsletter reporting his continuing contacts with his scientists, and he regularly published his own "four-star" rating of his various soil and water laboratories—a rating of great moment to the scientists involved.

A colleague of Raney's, Albert Grable, is also a close and judicious observer of current soil science research, reassuring scientists even as he too makes "suggestions" as to the kinds of projects that are likely to be funded in coming years (it is he who organizes and rationalizes the federal research budget for soil and water research). Other leadership roles are performed by research team leaders, department chairmen, and distinguished professors, as well as by leaders in other positions within the fraternity of specialists.

While such extended family ties can be functional, they can also insulate scientists from broader-based criteria for excellence. Agricultural scientists are not well represented in the leadership of high-status, science-wide organizations such as the National Academy of Science. This is in part because of their self-isolation, and in part because agricultural researchers may lack the poise cultivated by scientists in other settings. "They do not know how to present themselves," said one research leader. "They lack communication skills and breadth of understanding." But agricultural scientists have gained some positions within the scientific elite and are expanding this foothold. Integration is furthered by the desire of many nonagricultural scientists to become involved in food research.

Another source of increasing diversity in the field is the newest generation of agricultural scientists, whose formative experiences have been different from those described for the present establishment. Farm families are no longer likely to be isolated, and in any case farm kids no longer predominate among those enrolled in agricultural curriculums. Less than half of all "agriculture" freshmen at Iowa State University came from farm backgrounds.[21] Women now comprise about one-third of undergraduate agriculture enrollment, excluding the field of home economics, which was once the only appropriate "agriculture" subject for women. The women are coming—and ignoring the biases of some respected establishment spokesmen. C. B. Link of the University of Maryland, for example, feels it his duty to sound a warning: "I tell them there are jobs that women can do in horticulture, but they are very limited."[22] And Dr. Orville Bentley, dean of agriculture of Illinois and a leading spokesman for the experiment stations, has said of women veterinary students: "I told one of these women that it would scare me to death to have her work on my cows."[23] The women have not been deterred, however, and interest from this and other new student constituencies has sparked increased enrollment in agricultural subjects within the past fifteen years.[24]

Why are urban youth going into agriculture? "People today are more interested in the basics. It's the old back-to-nature and environment bug," said a University of Maryland senior. "Agriculture is the real world. Sociology, history, psychology and those things are all too abstract for me."[25] "It's all tied up with our concern with the environment and the world around us," explained an Iowa State University professor; "it's the desire to get back to nature, and agriculture is as close to nature as one gets."[26]

Changed student preferences have resulted in curriculum changes. The University of California at Berkeley has instituted a new program called "The Political Economics of Natural Resources" and another

48

program on integrated pest management. These curriculums have been oversubscribed, along with other new programs on conservation and natural resources, food and nutrition dietitics, and forestry. A Berkeley coordinator for agricultural students, Harold Heady, explains: " 'Conservation' is a buzzword, and 'managerial' things are attractive, because people are interested in the impact of science nowadays, and many of our students come through with the feeling that science may have done too much; therefore we need to manage it rather than produce it."[27] At Berkeley, which accepts only top academic performers from California schools, enrollment is low in the traditional preparatory fields for graduate work; entomology, plant pathology, and soil science, for example, had only ten to twenty majors each, as of 1977.

Many of the agricultural undergraduates will not, of course, go into research positions. A recent survey of fourteen agricultural universities indicated that 35 percent of the graduating bachelors took jobs in agribusiness and 20 percent went into farming, while the remaining 45 percent went on to graduate work, teaching, or extension services. According to John Pesek, chairman of the Department of Agronomy at Iowa State University, it would appear that scientific careers are not attracting some of the best students. "This year only a tenth of our students want to go on to graduate school, whereas we once had our pick of the crop," he notes. "Now many of the best students would prefer to go back to the farm."[28]

The new scientists will bring to the research establishment a different mix of backgrounds and motivations as they begin to replace an aging work force during the 1980s. One senior staff member is concerned about their number: "I don't kid myself. People over forty rarely produce anything. We depend upon the younger people, and we don't have many."[29] James Kendrick, head of the California-Berkeley experiment station, notes that he will be able to replace 80 percent of his scientists within a ten-year period due to retirements, and sees an opportunity in his choice of new people to redirect research to the extent he desires to do so.[30]

49

The changes that are coming in agricultural research are likely to be profound. Most of our agricultural scientists were nurtured, and now serve, within a subculture which is little studied and is not, perhaps, well understood by its many confident reformers. This agricultural subculture has been dependent upon its scientists, though not overly respectful of them, and the scientists have contributed mightily to agricultural modernization. It is the scientists' handiwork, most of all, which presently threatens this subculture and imposes a question mark upon the future of agricultural research. Where there was once a large pool of talented farm kids from which the research institutions could take first pick, there are now relatively few young farm people, and those few have at least as many attractive career opportunities as do nonfarm youngsters.

Fortunately, many nonfarm students are seeking careers in food research, and research leaders must now embrace the very imminent task of motivating and guiding a new scientific workforce recruited largely from outside their own subculture.

CHAPTER FOUR

Why Scientists Work

Although creative work emerges from a total cultural environment, some specific stimuli call it forth. For agricultural scientists these stimuli include rewards in the form of income, peer group esteem, freedom of choice, and a feeling of usefulness; resources in the form of research funding; and ideologies and group associations which they may wish to serve.

INCOME

Desire for a good income might seem to be a motive for choosing a scientific career, although in some respects salary may today be a negative incentive for agricultural scientists because of more lucrative possibilities in farming and agribusiness. I. Arnon, in his book on agricultural research organization, advocated low salaries for research "novices," to weed out those who were not truly interested in agricultural research.[1] Even for mature scientists high salaries may not provide a strong positive incentive. One federal research administrator complained that senior scientists do not feel obliged to work harder once they receive high salaries. "They just take them for granted," he said. Some studies have shown that careers in science chosen because of aspirations for income and organizational status are not as productive as are those pursued in an attempt to achieve self-actualization.[2]

Nevertheless, public research agencies place a high value on salaries as "incentive," both as a means for recruiting good people and as a reward for scientific achievement. There is embarrassment over the fact that many of the highest salaries have gone to research administrators rather than to scientists, and a deliberate effort is now being made to promote a larger number of distinguished scientists to the highest salary ranks. There is embarrassment, also, in the fact that private industry provides much larger starting salaries for some professions than do public research agencies.

PEER GROUP ESTEEM

Salaries, along with formal honors and awards (which may include financial awards), are, at best, tardy indicators of scientific status. The earlier and perhaps greater satisfaction comes through the publication of research findings that are regarded as significant by colleagues and research leaders. Recognition by one's peers—"peer approval"—may take the form of frequent citations of one's work by other scientists or of election to office in one's professional society. In the case of those with illustrious careers, there is the hope of gaining such cross-science honors as admission to the National Academy of Sciences, or even a Nobel Prize.

To reinforce peer approval, research administrators use quantitative measures of achievement. For example, Graham Purchase, as staff planner for federal poultry research, scrutinized computer tallies of publications and citations by federal research poultry scientists, and he distributed these tallies to the scientists in the field. A recent tally revealed that several poultry scientists had more than one hundred publications, while others had fewer than ten.

A scientist who does not produce much research faces peer pressure: "All about him his peers look down upon him," as one scientist put it. And this disapproval may translate into loss of research funding. The same scientist notes: "The formulas for dividing total dollars spent in research leads to the question—'Has that person contributed?' The peer pressure says, 'I'd like to have some of your resources if you're not using them any better.' " Besides losing funds and function, the unproductive scientist may suffer a relatively declining salary.

In the agricultural colleges, unproductive junior scientists may have to seek employment elsewhere, but senior scientists usually have tenure, a status in which dismissal amounts to a "cruel and unusual punishment" akin to the death penalty. One distinguished scientist explained the tendency not to fire scientists who have ceased to be productive: "In every institution, you come across cases where a man has outlived his usefulness, really, but for reasons of tenure and other things—an old friend—you can't tell him to get out and get another post. You lose too much in another way by grasping for efficiency. But large educational institutions pay a heavy price for inertia and slowness in order to get the benefits of academic freedom. How else should you do it?"

Indeed, when tenured professors have been hounded from their jobs the question of freedom has frequently been at issue. In a recent case, Robert Bradfield, a nutritionist, became *persona non grata* within the University of California–Berkeley agricultural extension service. As a result, he was obliged to keep a log of his activities for every hour of his working days, including nights and weekends, and was required to secure permission from his superiors before teaching any formal or informal classes, consulting or visiting his colleagues in the Department of Nutritional Science, doing any research, attending any meetings, using the university library, or contacting state or federal agencies. He was also deprived of his laboratory. In retaliation, Bradford pressed criminal charges against certain extension staffers who had opened his personal mail, searched his office, and sent university employees to watch his movements and listen to his conversations. After several years of conflict with his superiors, Bradfield finally resigned in 1978, under an agreement which reportedly involved a large financial payment in exchange for his dropping all charges against the extension service and his refusing to discuss the case in public.

The extension service had attempted to "fire" Bradfield on the ground that he did not do the job assigned him—originally, to participate in a program to improve the nutritional status of low-income members of minority groups. Bradfield, who had gained academic tenure as an associate professor in Berkeley's Department of Nutritional Sciences, appealed his dismissal to the university's tenure committee, which found no cause for dismissal action and therefore supported him.

Bradfield contended that he was being fired because he did indeed wish to study the diets of children of Chicano farmworkers and that extension officials feared such research would irritate their large grower clients. Bradfield surmised also that he had taken a too active role in seeking employment for minorities within the extension service. Bradfield's claim that he was doing things extension did not want done and extension's claim that he was not doing what extension preferred that he do may well be compatible. In any case, the action against Bradfield provided a rallying point for a political and media attack upon extension. The stress of fighting the establishment left Bradfield impoverished and in ill health, prior to the settlement.[3]

In the federal agencies, scientists have job security comparable to academic tenure once they have completed a short (nine-month) probation, although subsequently poor performers may occasionally be persuaded to resign, under pressure from colleagues and administrators.

Agricultural scientists are expected to show strong loyalty to their team or laboratory, their department, their agency—as well as their discipline—so that each can take personal pride in group achievements. Agronomists place on their cars bumper stickers which read, "Agronomy feeds the world." Iowa State Professor Wise Burroughs remarks, in pointed understatement, that his animal science department "has had a good name in the field, and we hope to keep it this way for quite some time to come."[4]

Agricultural scientists seeking status have still another reference group, the users of their research output, which Alex McCalla has called a second peer group. He explains: "Historically, substantial portions of applied agricultural research have been organized on a commodity basis and with strong relationships to commodity user groups. Many people in the agricultural research establishment have grown up with this association, so that to a considerable extent those inside the system share the same values as clientele groups. Therefore, they implicitly identify with their objectives."[5]

Since agricultural scientists specialize in studying a particular commodity, more often than not the commodity organization is likely to be close at hand, providing praise, financial and political support, and also criticism of research failures or of basic research that does not promise payoffs for local producers. An example of user recognition is that which was according to Dr. Donald Sunderman, a federal research agronomist stationed at an Aberdeen, Idaho, research center. Sunderman was pictured on the cover of the *Idaho Wheat Grower News*, and he was described in that magazine as being "largely responsible for the release of 11 new varieties which have brought Idaho Wheat Growers millions of dollars."[6]

THE "DILEMMA" OF
FREEDOM VERSUS USEFULNESS

Because a major incentive for scientific work is curiosity, scientists desire freedom to pursue research which interests them. Alex McCalla suggests that "the good scientist is a person whose personal curiosity is his or her greatest asset and therefore must first and foremost be interested in what he is doing."[7] And a scientist observes: "Young people go into agricultural research when they find out that they work in open space, and that they are less regimented at any level—you can do your own thing."

For scientists, freedom implies both the right to pursue an interesting subject *and* the "academic freedom" to report findings with integrity. As Don Price has noted: "Scientists . . . are manifestly concerned for the freedom of their own research, cherishing the *privilege* of unhampered investigation and teaching in academic institutions.[8]

Yet agricultural scientists are also taught to be useful. Most are, to some degree, "applied scientists" who are responsive to the demands of specific "users"—in part because the users are helpful to them in deciding research priorities, and in part because the users or "clients" support research funding. While the scientists' judgment as to what is useful may often be sounder than that of users, the fact remains that applied scientists must become interested in those questions the answers to which a user would find valuable. The scientists' curiosity must be tailored to fit the needs of the user. The head of a federal research center describes the process succinctly: "We go to the industry and we ask them, 'What's the biggest problem in genetic improvement?' and that's the project we undertake to do."

Industry spokesmen sometimes complain that agricultural scientists are, as Arnon suggested, more interested in "free, uncommitted research" than in working on practical problems,[9] and, therefore, that there is a tension between the interests of scientists and the needs of users. But most U.S. researchers are themselves utilitarians. Their satisfaction comes both from solving a problem and from seeing their solution used. "I was tickled pink," said soil scientist Al Grable, "that ranchers immediately used my mountain meadow forage findings." The frustration of engaging in useless research is registered in the lament of an old "utilization" scientist: "During World War II, we made some super-concentrated apple juice, snow-flake potatoes, and lots of things—we spent a hell of a lot of money on development of these things and they just sat there."

An example of rigorous commitment to "usefulness" comes from Randy Altman, who in 1977 was responsible for planning federal agricultural research on solar energy. At that time Altman was personally opposed to solar energy research because "the prices that would allow us to use it are not there yet, and may not be even by 1990. We want to do things that are useful. If it's uneconomic, it's 'ivory tower' to us."[10]

Thus, many agricultural scientists have resolved the dilemma of freedom versus usefulness by emphasizing usefulness. There are problems inherent in that solution. A science preoccupied with being useful may fail to notice the broader, secondary implications of its product, while researchers who relate closely to a particular set of users may find themselves avoiding research questions or soft-pedaling findings that would embarrass or disadvantage those users. Such researchers may even press the institution to stifle research by scholars in other fields which is unfavorable to their clients' interests; and to the extent that they succeed, the scientific institution becomes, incongruously, a censor.

A number of concerned agricultural science leaders have urged that scientists' training place more emphasis on the needs of the total society or the world, and become both more science oriented and, in general, more balanced in its treatment of freedom and usefulness. Glenn Pound, former dean of agriculture at the University of Wisconsin, has attempted to define the elements of balanced programs. Such training programs, he states,

> must give balanced exposure to the state of knowledge, the challenges of need, the most sophisticated methodology of investigation of the respective subject matter areas. They must give the student knowledge of the impact of related areas and leave him with an ability to place a problem within a broad perspective of science and to bring many approaches to bear on its resolution. Finally, the university environment should imbue a student with the reality of the usefulness of science, that science is useful when it is used. That is to say that there is a dimension of morality in science in that society's needs constitute the overriding priority.[11]

INCENTIVES THROUGH FUNDING

Researchers choose projects from among those research subjects for which funding is available. Younger scientists, at both federal and state levels, have to fit themselves into an ongoing research agenda. In federal agencies, scientists may have been hired to work with a specific research team whose leader negotiates with administrators and planners to mesh the team's interests with the tasks to which the agency is committed. Within the university experiment station, young researchers write research projects for which they receive "institutional" grants, which may provide the stipend for a research assistant, funds for minor expenses, and a portion of the researchers' salaries. The other portion of salary is covered under the university's teaching budget. To obtain additional funds for travel, equipment, and operating expenses, young scientists may join a team which is doing related research under outside funding.

Outside funding from governments may be in the form of competitive grants. Some of these are for specific problems and are awarded to one team among several which submitted proposals. Alteratively, a particular research subject may be designated, and evaluators may be able to fund all applicants who pose an interesting hypothesis and who appear to be able to carry through a sound research strategy. A number of agencies provide competitive grants for agricultural research, including the USDA, the National Institutes of Health, and the National Science Foundation.

In addition, various state and federal agencies, such as the Department of Energy, are ready to contract with agricultural scientists to gain information needed for development and implementation of their programs.

Finally, producer and industry groups, who are usually closely involved in decisions on institutional funding, may make direct grants or allocate funds received from processing taxes on particular commodities.

In short, researchers work diligently to obtain funding and often depend upon more than one source. At the same time, researchers may exercise some leverage over funding sources, in that they may be able to choose among several noninstitutional research sponsors and among a number of projects offered by each. Also, the scientists may find time to study additional questions that are not strictly a part of the projects for which they are being paid; much basic research, and many new research projects, are supported in this "piggyback" fashion.

It is the distinguished researchers at the distinguished institutions who have the most freedom in selecting projects. For example, Wise Burroughs, an Iowa State animal scientist who achieved distinction in developing diethylstilbestrol (which induced rapid weight gains in beef cattle), can afford to pick and choose: "We don't accept a grant unless we have a previous interest in the process. Now, if we were hungry for funds, we might be less selective. As it is, many others ask us for help that we do not give them."[12] Burroughs stressed the need for close contact with the beef industry, but it was clear that he, not the beef industry representatives, was the judge of relevance in the selection of research for his laboratory. As Harold Orlans has said: "It is probably true to say that the 'best' scientists can generally get support for the work they want to do; the 'average' scientists, for the work the government wants done."[13]

IDEOLOGY

Basic values may affect not only how hard a person will work but also for whom and with whom the work will be done. While each agricultural scientist may have a distinct personality, there are moral, social, and political values which tend to be shared, and these are influential. For example, it should be asserted at once that agricultural scientists are humanitarians. Most gain enormous satisfaction from the belief that their research has made it possible, up to now, to feed the world's increasing population. It distresses agricultural scientists to be rebuked for having caused temporary surpluses, or for having used the "wrong" technology for achieving food sufficiency.

Agricultural scientists, as a group, manifest distinctly conservative social and political perspectives. Government, though it is their employer, enters their group discussions as the overzealous regulator of pesticides and additives, as the busybody generating equal employment rules and other useless paperwork, as a misanthrope which pampers unproductive citizens with subsidies, using funds for "welfare" that would have yielded great returns if spent on research.

There are several possible reasons why most agricultural scientists are to be found on the right, or far right, side of the liberal-conservative spectrum. Don Price has said that scientists in general are conservative, even though they are themselves agents of change.[14] For agricultural

scientists, social background and associations provide additional explanations. In upbringing, most of the existing generations of agricultural
scientists shared a rural ethic which stressed work, uprightness, acumen, and independence. From the 4-H and Extension Service influences they learned about self-improvement through technological
progress. The agricultural colleges which most of them attended had a
commitment—often quite explicit—to reinforce patriotism, religion,
and conventional social wisdom. A recent Republican secretary of
agriculture, Earl Butz, reaffirmed these college values in a speech to
college administrators:

> Throughout history, land grant colleges and State universities have
> identified with the business and professional life in these States. You have
> taught the technology of production. You have functioned to create useful
> knowledge and then you have translated that knowledge in such a way as to
> train useful people—people who could produce for society and serve the
> needs of society. With some notable exceptions, the demonstrations and
> disruptions of recent years have sprung from universities and from colleges
> within universities in inverse ratio to the production-orientation of that
> school.[15]

As Butz noted, at some colleges the conservative impulse has been
challenged, particularly in recent years, by some students as well as
some faculty.

The colleges were one of several conservative reference groups with
which rural youth learned to identify. In most of rural America the
Republican Party was dominant until recently, and in the rural
Democratic South, elected officials kept their distance from liberal
Democratic presidents. As noted earlier, the American Farm Bureau
Federation, which became the grassroots contact for the colleges of
agriculture, also became aligned with the conservative coalition in
Congress and (after the New Deal) with Republican presidential ad-
ministrations.

Conservatism, imbibed during youth and training, was reinforced by
professional contacts. Research support in Congress came from conservatives on the appropriations committees; and researchers seem to

feel that they have fared better under Republican administrations. "We get something from both parties," said one research administrator; "the Republican administrations increase our budget and the Democratic administrations work us very hard." (The record indicates, as discussed in chapter 7, that Democrats have not been generous as often as Republicans, and that Democratic administrations have indeed frequently established new research missions, including marketing research.)

Also, researchers have worked with the efficient, better-endowed farmers—not the "disadvantaged" ones. They have worked with managers, not workers, and they have tended to share the managerial view that labor unions hinder "efficient resource utilization," a view which provides yet another incentive for reducing the amount of labor in agriculture.

The conservative orientation of agricultural scientists has been further reinforced by party and group conflicts over major farm issues in recent years, during which conservative politicians have advocated improving farmers' incomes by improving their efficiency, while liberal coalitions, including liberal Democrats, labor unions, and consumer advocacy groups, have attacked the "agribusiness middlemen" and the large farmers who have been the principal users of agricultural research.

The impact of these conflicts upon agricultural research policy is illustrated, in extreme form, by the fate of "marketing research" in recent years. This impact is outlined below through the perspective and activities of a federal marketing researcher, Dale Anderson. Anderson's research group tried to reduce food retailing costs by contributing to consecutive revolutions: first, the computer checkout system in grocery stores; and second, the preassembled mobile grocery shelf.

PEGGING THE VAMPIRE: LAYING TO REST
THE COMPUTER CHECKOUT SYSTEM

Dale Anderson, like former Secretary of Agriculture Earl Butz, is one of Purdue University's outspoken champions of the market economy

61

and the private corporation. Anderson, whose Ph.D. combined agricultural economics and industrial engineering, was brought into federal service by Butz, who was then (in 1953) an assistant to Secretary of Agriculture Ezra Taft Benson, also a Purdue graduate. Butz had previously advised Anderson, "Never work for the government," but then had persuaded him to ignore that advice as Butz himself had done.

At that time, the Benson administration was anxious to transform the Department of Agriculture into a helpmate of the market economy. Anderson went on to serve during the "golden years" of marketing research, the era which produced the supermarket. Then, as staff scientist for food marketing research, Anderson watched that research area decline. The decline began during the 1960s under Democratic Secretary Orville Freeman, whose assistant, Tom Hughes, slashed portions of the marketing research budget year after year. Congress invariably "restored" the funds, but only to the level of the previous year's allocations. The Democrats, with a perspective sympathetic to their farmer-labor agricultural coalition, were uninclined to help the middleman, who was presumably squeezing farmers, consumers, and workers. Studies were made and congressional hearings held on "profiteering" within the food industries, but these efforts failed to produce the expected evidence. Lack of evidence, however, did not deter another assault on food middlemen during the 1970s from consumerists such as Carol Tucker Foreman, executive director of the Consumer Federation of America. (Subsequently, both Foreman and Tom Hughes were given leading positions in President Carter's Department of Agriculture.)

Paradoxically, Dale Anderson's kind of marketing research had been authorized in the Agricultural Marketing Act of 1946, under a Democratic administration. The act offered to agribusiness research and extension services comparable to those that had been offered to farmers, specifically for the purposes of improving the marketing, handling, storage, transportation, and distribution of products.

"Basically, it has been our job to find better ways to handle products after they leave the farm," Anderson says.[16] "Better" has meant greater cost-efficiency, and reduction in the enormous loss or spoilage of food between the time it leaves the farm and the time it is eaten.[17] Public research under the 1946 act helped to create complex technologies for shipping "perishables," enabling residents of Chicago and Boston to enjoy, at reasonable prices, yearlong, store-wide supplies of impeccably fresh fruits and vegetables from California, Texas, Florida, and other distant places.

As Anderson views it, public marketing research has tended to prevent profiteering, by catalyzing innovation. He agrees with critics of the food industry that food corporations seek profits through exclusive markets rather than through efficiency, and that these firms therefore usually show a lack of enthusiasm for research that would require industry-wide revolutions in the technology of distribution and retailing. Anderson maintains that public research impelled the large firms to accept more efficient technology, and also reinvigorated the independent and cooperative wholesalers, saving them from being forced out by giant retailers.

In Anderson's opinion, the decline of market research is the chief explanation for increased food costs in recent years. "Look what happened to productivity per man hour when Secretary Freeman folded up marketing research," he says. By his calculations, market and product development research diminished from 30 percent of the total USDA research budget in 1965 to 23 percent in 1975.[18] Meanwhile, marketing costs increased by 60 percent between 1960 and 1970. The cost of food as a percentage of family income, which had steadily declined from 25 percent in 1935 to 15 percent in 1972, is now increasing.

And yet, as Anderson points out, the best opportunity for reducing food costs lies in making processing more efficient. "Remember, the farmer contributes only $36 billion to the annual food product, out of the total retail value of $194 billion. By shaving just a small percentage off the processing cost, we save so much money that policymakers are less apt to pat us on the back than they are to call us liars."

But critics have opposed "more efficient" technology which might increase food corporation power. They have delayed Anderson's dream—installation of the USDA-developed computerized checkout system for food stores, savings from which, according to a USDA technology assessment,[19] will be greater than the net earnings of the food chains that will use it. The system might also mean the elimination of many jobs in the grocery business. Under this system, the price of each item is listed on the shelf but is not necessarily marked on the can or package, all of which now bear computer marks for a scanner that would ring up the price. At the grocery counter there would be little to do but line up items for the scanner, sack them, and accept the consumer's payment. The principal savings would come from reduced labor costs and from the greater accuracy of the ring-up.

The prospect of a computerized checkout did not please the Retail Food Clerks International Union, which fired a regional officer who had cooperated with the pilot research project.

Consumer organizations had mixed reactions to the new process. The magazine *Consumer Reports* took note of the projected savings but decided, "We believe consumers should insist on price-marked packages."[20] If no prices were printed on the items brought home, consumers could not compare special grocery sale prices with the prices of goods on their own shelves, previously purchased.[21] Consumer advocates also worried that the initial costs of equipment might be passed on in the.form of higher prices, and that consumer privacy could be threatened if charge account purchases were passed, via the computer, to the firms that authorize credit.

On balance, most if not all consumer organizations opposed computerized checkout, but they also considered it inevitable. Not so for Carol Tucker Foreman, executive director of Consumers Federation of America. Mrs. Foreman's successful lobbying effort against the computer was described in a 1976 *New York Times* article: "Legislation to require that prices be kept on individual items was introduced and pressed by CFA . . . in numerous states and in Congress. The industry fought back, but by [1976] it had capitulated."[22] (More recently, how-

64

ever, the computer checkout system has begun to come into general use.) When the fact surfaced that Mrs. Foreman's husband, Jay, was administrative assistant to the president of the Retail Food Clerks Union, the *Wall Street Journal* called the lobbying effort "connubial consumerism."

The stifling of the computerized checkout also stalled the next big innovation, under which prestocked grocery counters would be wheeled out of a truck and onto the grocery floor; another enormous cost saving for the food chains, another big reduction in labor.

Dale Anderson believes, and the agricultural research establishment agrees, that consumers are best served by catalyzing innovation after innovation within the ranks of efficient producers, who are likely to be big farmers and big corporations. In contrast, Carol Foreman, who subsequently became assistant secretary of agriculture for consumer affairs, is more inclined to regulate middlemen than to subsidize public research for their benefit. Dale Anderson and his market researchers may survive, but given the liberal bias against helping "big business," they cannot expect growth under liberal administrations.

ASSOCIATION WITH OTHER GROUPS

Conservative values have enabled agricultural scientists to work in harmony with innovative farmers and agricultural businesses, and, despite their conservative values and their linkage with a conservative coalition, they have also gained some support from rural liberals. However, they have been slow in relating to members of a new coalition of liberal legislators and public interest advocates, including conservationists and environmentalists, consumerists, nutrition advocates, organic farmers, and others. These groups, sharply critical of research institutions, have achieved support for a new research agenda. Although the scientists approve of some of the new agenda objectives, such as soil conservation, they have manifested hostility toward these new groups, a hostility reinforced by some administrators and others within the institutions who serve as "mind guards." "Friends will criticize us if we go to conservationist meetings," says one scientist.

Radical critics of agriculture have watched university researchers take elaborate precautions not to be seen in serious conversation with them. The researchers are said to fear that a colleague might report such an incident to a superior, who would in turn give them a "black mark." Having refused to talk openly with public interest advocates, scientists cannot complain if these groups often have poor information, or use good information poorly. Recently, the scientists and public interest advocates have taken a second, less hostile, look at one another. Agricultural scientists, however conservative, have demonstrated that they are (as Bernice Eiduson said of scientists in general)[23] interested in antitheses, and research administrators have begun to recognize the value of good relations with new agenda advocates.

THE CONDITION OF AGRICULTURAL SCIENTISTS TODAY

Agricultural scientists, as a group, have been highly productive within (and perhaps because of) an intellectual environment constrained by rural upbringing, narrow education, and an imposed distance from the mainstreams of basic science research. External management of agricultural research has been pervasive. It has specified the size, location, and missions of research work sites. It has shaped a training curriculum which emphasizes applied research, and which creates needed specializations appropriate for solving current problems. Through background and training, most agricultural scientists share a common experience, as well as a conservative political ideology compatible with that of the farmers and agribusiness firms which have supported their private and public funding, and which use their findings.

Agricultural scientists, desiring mainly to be useful in improving the agricultural economy, have reason to be enormously satisfied with their performance. Beyond the satisfactions gained from personal and group achievement, they have received modest but good salaries, and the more productive among them have been honored by their professional societies and by client organizations.

Within this utilitarian environment, many scientists and administrators have nurtured a respect for scientific values and goals: they recognize the need for basic research, for developing methodologies, for preserving the integrity of research processes, and for replicating and freely disseminating research findings. These scientists have recognized, if clients often have not, that the pursuit of scientific values is indispensable not only in developing the "cathedral" of knowledge, but also in the production of useful information. Thus agricultural science has produced many distinguished scientists who have been able to gain independence for their own research, who have served as leaders in the development of their disciplines, and who have provided role models for the succeeding generation.

Conditions are now present for much change in the agricultural science community. The next generation of scientists, coming from more diverse backgrounds and embracing a variety of values, may pose some radical alternatives to existing production systems. The current generation of scientists is already moving to new agendas, and to the service of new clients who support these agendas. Meanwhile, there is increasing interaction with other science communities, including competition from other sectors for research grants on subjects of great interest to agricultural scientists.

Signals from the political environment, once consonant and clear, are now mixed and confusing. Agricultural scientists, and even more so their administrators, are pressed to adapt in ways that will preserve the strengths of agricultural research while opening doors to new incentives, new missions, new support. In the following chapters we will discuss the confusing new political environment.

CHAPTER FIVE

Research Administration

Research administrators have conformed, perhaps too well, to the triangle model of subsystem power. As later chapters will show, they have interacted well with leaders of industry groups and with key legislative committees, and it is reasonable to assume that they were selected and trained to do so. Today, however, these administrators are challenged to relate to a much larger constituency.

THE RESEARCH ADMINISTRATORS

Who are the administrators? Presumably their behavior reflects, to some extent, their origins and training, and these are examined for five sets of administrators in tables 5.1 through 5.7. Line administrators comprise three sets: first, the Agricultural Research Service (ARS), including persons who served within the central administration, in regional or area offices, and as heads of major research installations; second, the deans of state colleges of agriculture and directors of the experiment stations; and third, the state university presidents. (The presidents have taken a hand in the development of the agricultural colleges, and some are still influential in their formal role as chief administrative officers for agricultural research, extension, and teaching.)

The two staff groups analyzed here include the Cooperative States Research Service and the national planning and program analysis staffs of the Agricultural Research Service.

Also presented in the tables are characteristics of the membership of two National Academy of Science evaluating committees, whose evaluations of research leadership will be discussed in this chapter.

Tables 5.1 and 5.2 reveal that research administrators are drawn largely from the ranks of agricultural scientists. Virtually all experiment station directors and Cooperative States Research Service (CSRS) staff are Ph.D's, as are a large majority of ARS line and staff officials, and most of their degrees are in natural science fields. (The major exception is that seven of the sixty-nine state deans and directors are agricultural economists.) Well-represented disciplines include agronomy, soil science, animal science, veterinary medicine, plant pathology, entomology, genetics, and nutrition.

TABLE 5.1: Comparing Science Leaders: Percentage with a PH.D. Degree

Scientist Groups in 1977	Total Identified	Total Found[a]	Received Ph.D.
	Number	*Number*	*Percentage*
Experiment Stations:			
CSRS staff	38	33	97
University Presidents	54	51	80
Deans, Directors	84	69	97
Agricultural Research Service:			
Area-Regional-National Administrators	53	40	80
Planning (NPS, PACS)	64	41	83
Science Evaluators:			
Pound Committee	50	46	98
World Food and Nutrition Study	150	119	86

[a] Information was sought from the following bibliographical references: *American Men and Women of Science*, vols. 12 and 13; Marquis *Who's Who* publications; and directories of the *American Journal of Agricultural Economics* and the American Economics Association.

TABLE 5.2: Comparing Science Leaders: Field of Specialization

| | Research Administrators | | | | | | Research Evaluators | |
| | ARS | | Land Grant Univ. | | | | | |
	Admin.	Planning (NPS, PACS)	Univ. Pres.	CSRS	Deans Dirs.	Total	Pound Comm.	WFNS
Agronomy	7	5	1	4	8	25	4	11
Soil sci.	5	4	—	—	8	17	4	11
Animal sci.	—	2	—	3	7	12	3	4
Dairy sci.	2	—	—	1	3	6	—	2
Poultry sci.	—	1	—	2	—	3	1	—
Horticulture	—	—	—	—	5	5	1	1
Food sci. and technol.	—	—	—	2	3	5	2	1
Home econ.	—	—	—	—	—	—	—	—
Ag engineering	4	4	—	2	1	11	—	4
Veterinary med.	2	4	2	2	2	12	1	8
Plant pathology	4	—	1	1	6	12	1	—
Plant physiology	—	2	—	1	1	4	2	12
Physiology, human and animal	—	1	1	1	1	4	—	—
Microbiology	—	—	—	2	—	2	2	2
Ecology	1	—	1	—	—	2	2	7
Entomology	3	6	—	2	1	12	1	5
Genetics	2	3	1	1	5	12	4	9
Nutrition, scientific (excludes Nutrition in home econ. and dietetics)	—	1	1	4	6	12	3	9
Physical sci.	5	1	3	1	1	11	3	10
Ag. econ.	1	1	1	1	7	11	1	10
Other soc. sci.	—	—	1	2	1	4	10	48
Business management	—	—	—	—	—	—	—	18
Other	5	4	10	8	2	29	7	28

Administrators have many similarities, and a few differences. Table 5.3 indicates that more than half of the state deans and directors are from farms or very small towns, as are a large minority of federal line and staff. However, at least one-fifth in all groups were born in a large city. A large proportion of the academic degrees are from colleges in

70

TABLE 5.3: Comparing Science Leaders: Size of Birthplace (1970)

Scientist Groups in 1977	50,000 or more	5,000 to 50,000	Farm or less than 5,000	Total
Experiment Stations:				
CSRS Staff	21%	18%	61%	100%
Univ. Pres.	35	18	47	100
Deans, Dir.	25	22	53	100
ARS:				
Area-Region-National				
Admin. Planning (NPS,	28	28	45	101
PACS)	29	27	44	100
Science Evaluators:				
Pound Comm.	29	27	45	101
WFNS	47	18	35	100

TABLE 5.4: Comparing Science Leaders: Region of higher education degree

Scientist Groups in 1977	West	Midwest-Plains	East	South	Outside	Total
Experiment Stations:						
CSRS Staff	19%	47%	13%	19%	1%	99%
Univ. Pres.	23	30	25	22		100
Deans, Dir.	15	42	22	21		100
ARS:						
Area-Regional-National Admin.	15	39	17	25	4	100
Planning (NPS, PACS)	10	41	26	17	6	100
Science Evaluators:						
Pound Comm.	7	54	19	17	3	100
WFNS	15	42	28	9	6	100

the Plains and Midwest (table 5.4), understandably, since most of the large graduate colleges are located there.

Table 5.5 indicates that the great majority of administrators are from generations which came of age before or during World War II.

Table 5.6, an indicator of breadth of training, suggests that some

71

TABLE 5.5: Comparing Science Leaders: Birthdate

Scientist Groups in 1977	Before 1910	1910– 1919	1920– 1929	1930– 1939	1940– 1949	
Experiment Stations:						
CSRS staff		21%	70%	9%		100%
Univ. Pres.	2%	37	49	12		100
Deans, Dirs.		35	51	14		100
ARS:						
Area-Region-National Admin.		30	56	13		99
Planning (NPS, PACS)	2	34	41	22		100
Science Evaluators:						
Pound Comm.	7	39	43	11		100
WFNS	5	24	37	22	11	98

TABLE 5.6: Comparing Science Leaders: Ivy League and Land-Grant College Degrees

Scientist Groups in 1977	Percent who received one or more academic degrees from:	
	Land-Grant	Ivy League & Chicago[a]
Experiment Stations		
CSRS Staff	97%	0%
Univ. Pres.	59	22
Deans, Dir.	99	6
ARS:		
Area-Regional-National Admin.	83	0
Planning (NPS, PACS)	83	15
Science Evaluators		
Pound Comm.	87	4
WFNS	76	24

[a] Some science leaders received degrees from both categories and so are counted in each.

groups have rather diverse educational backgrounds—particularly the university presidents, the World Food and Nutrition Study members, and the federal planning staff. State deans and CSRS staff are uniformly the products of the land-grant system which they now serve, but it should be noted that the ranks of ARS planning staff do include 15 percent without land-grant experience, some of whom were educated in foreign universities.

Table 5.7, which is an effort to show breadth of professional interest, indicates that most administrators and planners have membership in an agricultural discipline, and most also have ties to a general discipline. More than one-third belong to a science-wide association such as the American Association for the Advancement of Science. Membership in a science-wide association may indicate an interest in its broad orientation, but it is not known whether scientists gain breadth as a result of membership. It is similarly difficult to judge breadth resulting from travel. Agricultural science leaders may be as well traveled as military officers and members of Congress, because their skills are in demand throughout the world.

Land-grant university presidents, compared with agricultural science leaders, are less likely to have farm / small-town backgrounds or to

TABLE 5.7: Comparing Science Leaders:
Breadth of Professional Membership[a]

Scientist Groups in 1977	Agricultural Subdiscipline	General Discipline	Science-Wide
Experiment Stations			
CSRS Staff	78%	66%	41%
Univ. Pres.	23	90	33
Deans, Dir.	73	68	44
ARS:			
Planning (NPS, PACS)	76	80	43
Area-Regional-National Admin.	74	63	40
Science Evaluators			
Pound Comm.	50	100	63
WFNS	48	79	49

[a] Designation based on *Encyclopedia of Professional Organizations*. Data are for those whose biographies were found in one of the biographical directories listed in table 5.1.

be from the Midwest. Almost half were not educated at land-grant schools. They come from many career fields, and 20 percent of them do not have the Ph.D. As indicated in table 5.7, about one-fourth of all land-grant presidents associate themselves with an agricultural subdiscipline.

THE JOB OF RESEARCH ADMINISTRATOR

At least three demands are made upon research administrators: to develop objectives for research; to elicit political support that can be converted into resources; and to coordinate and manage these resources to achieve the chosen objectives. Political support derives, to some extent, from the institution's choice of objectives and its reputation for achieving them.

Developing Objectives

A large planning apparatus has been created within research institutions, to enunciate programs under which all research projects are aggregated and justified and according to which future projects are started.

Is this a useful function? Should research be bureaucratically programmed? No, answers Michael Polanyi, arguing that science more capably organizes itself. The research process, as Polanyi describes it, is a series of independent initiatives, mutually adjusting, which result in stages of discovery. Polanyi believes that "any authority which would undertake to direct the work of a scientist centrally would bring the progress of science virtually to a standstill."[1] Many scientists no doubt cheer Polanyi's stand, but, in reality, agricultural research has long been politically directed, as we shall see.

Another guiding hand offered as preferable to the bureaucratic one is the technological marketplace, to which researchers are said to be quite responsive.[2] Comparing research systems in the U.S. and Japan, Hayami and Ruttan surmise that Japanese technology developed in response to the existence of abundant labor there, while the U.S. developed labor-saving technology in response to the abundance of

74

land in this country.[3] Three points should be considered here, however. First, technology has often defined the supply of agricultural resources, rather than the other way around, as when dryland farming techniques opened large areas of the world to cultivation, and when mechanical cotton pickers displaced several hundred thousand sharecroppers from rural areas in the South.[4] Second, market cues may be inappropriately short-range, as in the current selection of energy technologies. And third, markets ignore so many "externalities"—such as environmental contamination—that their influence should often be resisted.

Those who think that economic stimuli are good guides for technological innovation may use administrative objective-setting in a complementary role. "Feasibility studies" and other mechanisms can specify the needs of the technological marketplace.

Administrators have a large role under still another theory: that "society" (rather than just the economy) has "goals," on the basis of which public resources are allocated. Within this framework, administrators who would seek public resources must identify and serve these social goals.[5] Again, there are points to be made in response to this theory. Political scientists and politicians recognize that effective support and opposition come not from the whole of "society" but from specific small sectors. Even the occasional wide or "diffuse" support for research provided through the ballot box is usually elicited from rather specific publics such as farmers, conservationists, and environmentalists. Close observers of political behavior are likely to regard references to "society's goals" as naive or self-serving.

Yet there is reason for administrators to try to anticipate future political demands, simply because the research begun today yields its return in a subsequent political environment. In a sense, the administrator keeps two books of "objectives"—one of which rationalizes ongoing research in terms of its relevance to current political demands, while the other sets research directions that future administrators will be able to rationalize in the light of future demands. Especially in light of the fact that the research chiefs of the future are serving now as scientists and deputy administrators, the research organization has reason to be future-oriented in choosing its objectives.

Developing Political Support

In recent years, the constituencies interested in agricultural research policy have expanded. Previously, research administrators had become accustomed to working closely with clients within the agricultural industry and with a handful of legislators on federal and state legislative appropriations committees. They had also cultivated the support of their nominal superiors within the Department of Agriculture.

During the past two decades, many legislative committees have extended their oversight to one or another aspect of agricultural research. The situation is similar within the executive branch, where nonagricultural agencies such as the Department of State, the Environmental Protection Agency, the Food and Drug Administration, and the Department of Energy have exhibited a strong interest in directing agricultural research.

There are numerous new constituencies which are pressuring research administrators to relate to their interests. Environmentalist groups, for example, have criticized pesticide research, and have advocated a shift to other means of pest control. Administrators are now challenged to develop a mix of pest research programs that will gain support from environmentalists (or reduce their criticism) without losing the support of industry clients. There are other interested publics as well, including not only environmentalists, but also various humanitarian elites. To shift to research objectives which please all these constituents, administrators must generate new motivation and expertise within the institution, while warding off hostility from scientists who would prefer to go on doing traditional pesticide research. The task of developing new constituencies into supportive clienteles is a demanding one.

When administrators say, as in the Pound report, that they seek to fulfill "society's needs," presumably they have in mind a set of missions that can attract such diffuse public support. Diffuse support has usually been obtained from the promise rather than the fulfillment of these missions, as evidenced by support given to the early scientist-entrepreneurs and by the apparent absence of supportive publics during the past two decades, when research had clearly earned major credit for cheap food. Recently, a few agricultural research administrators, such as Sylvan Wittwer, director of the Michigan Experiment Station, have shown that administrators, no less than early-day scientist advocates, can cultivate a supportive public based on the promise of miracle biology to meet world food and energy needs.

Managing Resources to Achieve Objectives

Obviously, the task of implementation is interdependent with those of defining objectives and eliciting political support. The objectives chosen must attract clientele support and / or diffuse support, and must also attract those who will do the work of research. The administrator's integrity is at stake, both inside and outside the agency, if he or she seems to promise more than is likely to be fulfilled.

The administrator's contributions to fulfilling promises lie in maintaining the organization and in coordinating its activities. Maintenance functions include recruiting and motivating able researchers, because research is a highly "labor-intensive" activity. There are manifold housekeeping tasks—among them, supplying equipment, providing a satisfactory work environment, and generally minimizing the psychological and physical burdens of doing research. There is also the task of maintaining boundaries: in interacting with other agencies, the administrator seeks not only a favorable exchange but also the maintenance or expansion of the agency's jurisdictional boundary lines.

As another form of organizational maintenance, the administrator must justify the organization's output, which is really a product of earlier decisions, in terms of expectations emanating from the current political environment.

EVALUATIONS OF RESEARCH ADMINISTRATION

How well has research administration performed the functions of setting objectives, developing political support, and management? There have been evaluations from various perspectives, with various findings. When research administrators are evaluated on the basis of the returns from investment in research, as corporation executives are evaluated by the profits they produce, performance must be rated superior. Cost-benefit studies show rates of return from all kinds of agricultural research which run as high as 90 percent per year.[6] Economist Kenneth Boulding, who once passed out academic grades to sectors of the agricultural economy, gave research institutions the only "A," because of their success in stimulating growth and development. Boulding added, "We might even make it an A plus plus."[7] In the light of this achievement, research administrations have been inclined to say, "We must be doing something right." Additional proof is needed, however, because such high returns are characteristic of agricultural research institutions elsewhere in the world.

Indeed, recent evaluators of research administration have found much to criticize. An in-house evaluation by the 1972 Pound committee found managers to be sometimes complacent.[8] Administrators have been criticized for pursuing efficiency at any cost, like Casey Jones at the throttle: in their rush toward productivity, they are said to have failed to replenish the fund of basic knowledge and to have ignored the undesirable side effects of new technology. They have also been criticized both for failure to muster adequate political support from within the agricultural industry and for failure to develop new sources of political support.

The Pound committee noted that in previous decades both the structure of agriculture and the organization of scientific inquiry had changed in fundamental ways. The report challenged agricultural research institutions to become more responsive to these changes. Although the Pound committee reported "many programs of excellence" and "many new program thrusts and organizational changes designed to cope with new problems," these were seen to be exceptions to the general pattern. The committee found too little attention being given to basic agricultural research; poor coordination and integration

78

of federal and state research; and unbalanced commitments of resources.

The committee's report also pinpointed an even more crucial problem:

in the area of management of scientists the committee found very disturbing evidence of ineptness with direct impact on research quality. Administrative structures and philosophy were found that reduced the decision-making power and freedom of movement of the scientists, with repressive effects on the quality of science. Programs for staff improvement, throughout the system, are either grossly inadequate, or ineffectively used, resulting in stagnation and premature obsolescence of the scientists.[9]

The Pound committee was severe in its criticism, by comparison with subsequent scientist evaluators, even though it was an in-house group in terms of the birthplaces and education of members (see tables 5.3 and 5.6). Its members did, however, have a strong orientation toward the general, rather than agricultural, disciplines (table 5.7). The committee had been created by Secretary of Agriculture Clifford Hardin, himself an agricultural economist and a former president of the University of Nebraska. It was chaired by Glenn Pound, dean of the School of Agriculture and director of the Wisconsin Experiment Station. A possible explanation for the severity of this committee's report is that its members may have been selected as individuals with a capacity to exercise a critical or enlightened perspective. It is also probable that the Pound committee did not anticipate that its blunt language, intended to achieve reforms, would be reinforced by more flamboyant analyses, such as Jim Hightower's, and then be used to support a conclusion that existing research administration was biased and incompetent.[10]

Two other scientist evaluations, following in 1975, were sponsored by the Board of Agriculture and Renewable Resources, National Academy of Sciences (NAS). These emphasized the need for "more resources" rather than for leadership change. But both reports called for new directions, and particularly for more "basic research."[11]

These recommendations were supplemented in 1977 by those of the World Food and Nutrition Study, requested by President Ford, whose task was to recommend new research directions to achieve the mission of adequate food for the world.[12]

The World Food and Nutrition Study suggested a number of institutional changes, including the appointment of an assistant secretary of agriculture for research and the strengthening of the interdepartmental committees which coordinate agricultural research with many missions. This group also recommended a large "competitive grants" program under which funds would be awarded by scientist peer panels rather than by research administrators, as a way to induce a rapid shift toward new objectives, to improve research quality, and to enable scientists outside the agricultural research institutions to become involved in food research. Competitive grants were recommended also in influential reports sponsored by the congressional Office of Technology Assessment,[13] and by two subcommittees of the U.S. House Committee on Science and Technology.

Although these various evaluating committees were positive and supportive in their comments, perhaps to avoid angering administrators, they asked for broader, more effective leadership. The competitive grants proposal urged by these groups, in particular, was viewed by many research administrators as a challenge to their decision-making role and as a vote of no confidence in existing institutions. The House Science and Technology Committee report, issued following extensive hearings, stated that witnesses had frequently reiterated the Pound report's conclusion that "most of agricultural research is outmoded, pedestrian, and inefficient."[14]

These subcommittees reported that "agricultural researchers need the guidance which only a national policy can provide." Granting that the existing agricultural research institutions were the possessors of the best "resource base" for modernized agricultural research missions, the report stressed that research leadership had yet to exercise its responsibility "to discuss alternative strategies and the elements of constructive criticism from within and outside the system."[15]

Major criticism had indeed come from public advocacy groups, which accused institutional leaders of indifference, even hostility, toward groups and government agencies concerned with environmental protection, civil rights, farm worker security, food safety, and other missions impacted by agricultural research. Critic Jim Hightower concluded: "The present decision-making apparatus is hardly conducive to the massive changes that must occur. That apparatus must be shaken, and other interests must become dominant in shaping land grant research policy."[16] One reformist congressional staff member who was instrumental in enacting a provision for solar energy research for agriculture quickly decided after a visit to the appropriate federal research administrators that "they are against us."

A different criticism, from those who tried to coordinate agricultural research institutions, was that many administrators were reluctant to work for common goals. When the 1977 research charter was being shaped in Congress, state research and extension leaders were split on such issues as whether and how to allocate funds for the "1890s" (predominantly black) agricultural colleges. "You can't believe the disunity in this group," said a 1979 coalition leader. "The extension director and the station director will smile at one another at the barbeque, then go in and try to cut into the other's budget."

One agricultural experiment station director charged that leadership was relatively inactive. Referring to the national coordinating office for experiment stations, the Cooperative States Research Service, he said: "You must have noticed that they have been sitting around there for years doing nothing. There are many things they could be doing—communicating with other scientific establishments, and planning larger funding." About leadership of the federal Agricultural Research Service he concluded: "ARS is top-heavy with administration—regional leaders and the program people in Washington. You would have to tell me what they do." And about state experiment station leadership: "You must know that the leadership of agricultural research is bad. Nobody does anything."

These various criticisms of contemporary research leadership can be met with at least three defenses, the first being that behavior has changed in some cases. One young experiment station director, for example, now runs an open, active administration: "If you immerse yourself in the dialogue," he says, "you begin to see patterns of need." A second defense is that federal research agencies *have* restructured themselves around national missions, in 1952, 1972, and 1977; one can even argue that they tried too hard.

The third point to be made is that some critics may have unrealistic expectations: They appear to believe that research objectives can be neatly stated and quickly changed, and that researchers can be tightly coordinated without sacrificing creativity. However, research administrators may themselves hold one of the least realistic expectations: that it is possible to initiate major changes in direction in the absence of pressure from critics.

MECHANISMS FOR ORGANIZING & JUSTIFYING RESEARCH

The effort to organize research into programs is a relatively recent one. Typically, research decision making has been decentralized and disaggregated. At the state experiment stations, research proposals are submitted by individual scientists following their own preferences as informed by knowledge of their scientific fields and by some contacts with the industry and with departmental colleagues. Then, in the stronger academic departments, the department head reviews these proposals for soundness and relevance. Following this, the experiment station director reviews the proposals and sends them on to the national office of experiment stations for perfunctory approval. Until recently, many proposals tended to be carelessly—or craftily—vague. Effective supervision has been indirect, in the form of decisions as to which kinds of research specialists should be hired and which individuals should be hired, and of annual judgments as to how much money to give to each academic department. Research direction has been much influenced through informal channels, including pressures from legislators and agricultural groups.

Within the federal Agricultural Research Service (ARS), between 1952 and 1972, research on each commodity-oriented topic, such as on "diseases of sheep," was headed up by an investigation leader who coordinated the research of several teams at different locations. This investigation leader exercised strong control through budgets and personnel transfers, and served as a point of effective access for legislators and farm group leaders who wished to influence research decisions.

This disaggregated decision making at state and national levels was the object of recent reforms. The first step in reform was a joint experiment station–federal research task force, formed by Secretary of Agriculture Orville Freeman in 1965 in response to the request of a Senate committee. This joint task force painstakingly reviewed the thousands of state and national research projects, developing sets of planning categories into which all projects could be fitted.[17] Thus was begun the Current Research Information System (CRIS), in which each research project is expected to have an accurate and meaningful description. Projects are described by "key words," by which they are aggregated and retrieved by computer.

The CRIS system is not perfected. A 1977 General Accounting Office study pointed out that, partly because the CRIS format permits only 300 words in project descriptions, the stated objectives are often vague, there may be no description of linkage with other research, and technical methodology is not fully stated.[18] But even with this imperfect information base it was possible to organize research according to mission. In 1972 a national planning staff of about sixty senior scientists was created within the Agricultural Research Service. A pyramidal program structure which embraced all ARS research was developed. At the top level of this planning pyramid, research was categorized according to the major missions of the U.S. Department of Agriculture. These missions were subdivided conceptually into research "goals," with one or more for each mission. At the bottom of the pyramid were more than three thousand ARS research projects. Among the functional intermediate categories were approximately fourteen "national programs" which served as the "decision packages" or major aggregations used in formulating the budget. These programs were further divided into sixty-seven "national research programs" which served as units for national research planning, as well as several "special research programs" which focused and mobilized research for relatively short-range objectives. For each of these sixty-seven national research programs there was a written document which outlined the "visualized" research plans, and in some cases also contained updated summaries of completed research. Each of these elaborated national research programs was considered to be a "consensus" document, "because everybody in research positions was involved in writing it and it reflects what we are doing."[19] In theory, this was the bible from which researchers were to be guided.

Each national program was supervised by a national program staff member who could recommend new projects to be undertaken and existing ones to be redirected or terminated, with consequent shifts in personnel and funding. National planning staff members, operating both as individuals and as functional subgroups, reported to a national associate administrator. They were expected to work with line administrators at all levels in improving and redirecting research, but scientists whom they supervised expected these staff members to "hold the line" against shifts of resources from their functional areas.

Conflicting expectations appeared, as well, in several of the devices for making planning more objective, better informed, and consensual. One of these devices was the "task force," which would aggregate the wisdom of scientists and administrators in different agencies and disciplines and among clientele groups to deal with a problem in which all had a stake. Procedures were devised for the task forces which would elicit full discussion and would then allow the group to reach a collective judgment.[20] Group decision making was also used for framing agency budgets. But in the last analysis, the members of the task forces were more likely to see themselves as advocates than as judges, casting their arguments and votes in favor of their own institutional interests.

Various quantitative tools have been used to improve objectivity, but these too produce a limited or piecemeal view. "Feasibility studies" are made, for example, to indicate which kinds of research are likely to have a "take" in the marketplace, but feasibility is only one of several potential criteria; typically ignored, for example, has been the question of whether or not a new technology would contribute to the production of nutritious and safe food.

Cost-benefit studies have been used to pick the research projects which would provide the best return on investment. Such studies have admitted shortcomings: there are methodological problems in measuring changes in productivity due to research;[21] many costs are usually not considered, either inadvertently or because they are not measurable;[22] and it is difficult to estimate the value of "second-generation" costs and benefits, including those which spill over into other countries or into the future. The shortcomings of cost-benefit studies are so great that a former director of USDA research, Ned Bayley, despaired of using cost-benefit studies to compare diverse research areas, or to compare investment in research with investment in action programs.[23] Sponsors of a recent conference attended by leading analysts perceived a consensus among analysts that "there are severe limits on our ability to make quantitative objective assessments of the value of particular kinds of research. This implies that choice must continue to be determined to a great extent by subjective judgment. The objective criteria available to guide research decision-makers through the uncertainty surrounding research decision are limited."[24]

Despite disclaimers, cost-benefit and feasibility studies continue to be used, and the result may be systematically to neglect research which lacks immediate relevance or payoff. As an example, one research administrator told a congressional committee in 1975 that research to derive protein from nonagricultural sources had been calculated to be economically unfeasible. He argued that such food would cost twice as much as that from existing foods. A committee member retorted: "May I comment on that?... We are devoting tens of millions of dollars to research in generating electricity from photovoltaic cells with a present cost of 100 times what it is for other sources.... When you get down to only twice you're well within the range of an effort which ought to be exploited."[25]

The congressman's example was a good one. The technology to which he referred would extract electricity from the sun's rays, and with further development it has since become economically feasible for some uses.

The use of feasibility studies probably explains (or at least rationalizes) the reluctance of agricultural research agencies to investigate new ideas. As scientist Charles Lewis has pointed out, it is uniquely the role of public research agencies to engage in both basic research, the usefulness of which is not yet even an issue, and developmental research, at which stage "nobody can predict accurately...and many things do not pan out." Lewis notes: "Once you have evidence that it will pay off, evidence of high frequency payoff, then private companies can do this and they should invest."[26]

RECONCILING PLANNING WITH
ORGANIZATIONAL & POLITICAL REALITY

Lofty national research missions have to be reconciled with mundane human and organizational needs. For example, in the case of twelve projects selected at random in a category of "high-priority" research, it turned out that there were workaday reasons for selection in eight cases: to compensate a university for the federal agency's having pulled out of a joint research project; to break in a new Ph.D.; to give work experience to some new people; to shift an inadequate administrator into a research job; to use "extramural" funds which had become available from another agency; and to provide a job for two scientists whose project was being closed down. Two of the projects were obsolete, and though one of them was being closed down, the other was being continued to avoid confrontation with its obstreperous leader.

Game playing occurs, chiefly in order to meet the presumption that research can be planned in short time frames and oriented to national missions. Game playing, an activity traditionally undertaken in behalf of basic research and of scientists' autonomy, is also used to blunt the thrust of new missions. Game-playing routines include the retitling of projects as a way of assigning them to desired missions. And, within budgets, "we always give the Assistant Secretary what he wants up front," in the hope that the agency can have its way on the bulk of the budget. But this "up-front" strategy is difficult to pursue, now that the required format exposes the major features of the budget.

Agencies, on the one hand protective of ongoing research, on the other hand look upon new priorities as opportunities for adding "lines" in their budgets. Another game, "crest riding" (formulating proposals which cater to momentary enthusiasms), is played with finesse by agencies, by academic disciplines, and by individual researchers. But the research community places some constraints on "crest riding," from a sense of academic honesty. Many scientists are distressed by the use of popular "buzz words" to characterize existing research, and by unrealistic promises of quick "solutions," both of which give advantage to those agencies and individuals who are willing to indulge in "puffery."

COMMENTARY

The earnest effort to improve research administration has produced a set of Catch 22's. The CRIS system, for example, has increased awareness of research being done, but the sharing of this information has produced jealousy as well as cooperation, especially during years of declining budgets. Critics of agricultural research, including some within the secretary's office and within the White House, have used CRIS information to justify reducing funding for research. While planning staffs have proven helpful in mobilizing multi-disciplinary research, they have also been available to rationalize the status quo. Structural reforms have irritated rural legislators and agricultural groups (as indicated in the succeeding chapters), as well as the research scientists themselves. Quantitative measures using the narrow criteria of industry acceptance and profits have been far less useful as a means of evaluating research than scientific intuition and political judgment.

A broad reorganization of agricultural research was attempted by the Carter administration. A new Science and Education Administration, created by Secretary of Agriculture Bergland's office, sought to coordinate federal and state agricultural research, education, and extension agencies as it moved them to new missions. Two permanent advisory committees were established: a joint council which provided representation for agricultural research institutions as well as for other potential "producers" of agricultural research; and a users' advisory committee which cultivated support from such potential clienteles as consumers, environmentalists, and nonagricultural agencies of government. A staff was created to develop research programs for such alternative technologies as organic farming and integrated pest management. This new structure was abolished by Secretary of Agriculture John Block.

It appears that agricultural research administrators, as a group, found it difficult to tolerate the dissonance and uncertainty of this broader framework. They were always apprehensive about competitive grants, which added funds to the total research budget but also added new missions. In 1979, the lobbyist retained by the state experiment stations and extension services, Ed Jaenke, persuaded the House appropriations subcommittee to increase institutional funding at the expense of the administration's budget for competitive grants. Thus, administrators opted for the status quo, even while both agriculture and science were in a process of transformation.

The administrators' lobbyist, Ed Jaenke, seems confident that there is a "sleeping giant" to be wakened within the agricultural establishment itself, and that, on the other hand, critics of agricultural research will not be a menace much longer. New people within the administration, Jaenke says, are ready to "ride herd on critics from within, and the public interest groups will be losing interest in agricultural research as other glamorous crises loom up."[27]

Many administrators remain confident that the commercial sector needs them too much to let them down. "Commercial agriculture has used up all the technology," says a national staff planner. "They are desperate for new research, and so they will support us." There is some faith, too, though little evidence, that consumers reacting to rising food prices will support research. However, friends in Congress have advised research leaders otherwise. Even Congressman Jamie Whitten, a major broker of industry power, has indicated that they must appeal to a wider constituency.

The impulse to return to an industry clientele is probably attractive to a majority of agricultural administrators and researchers. For them this would be "business as usual," in line with their own values, specialties, and associations. Meanwhile, it is not clear where decisions within research institutions are now to be made. Will the decision process reside in the planning mechanisms and the new coordinating structure? Or will these be bypassed by research chiefs and their lobbyists working directly with agricultural groups and rural legislators?

CHAPTER SIX

Industry Groups

The private groups which make up one segment in the triangle of agricultural subsystem power can be classified within six categories: general farm organizations, commodity organizations, agricultural trade associations, farmers' cooperatives, agribusiness corporations, and agency groups associated with a particular program, such as the Association of Soil Conservation Districts. Many of the groups have characteristics which qualify them for inclusion in more than one category, as, for example, the American Farm Bureau Federation (AFBF), which is considered a general farm organization, though it has organized a network of cooperatives has provided marketing services for producers of particular commodities. Each group has been classified here according to its predominant image.

Table 6.1 offers a perspective on the number of industry groups of each type which have been active in agricultural decisions. The first column of table 6.1 lists the membership of a 1979 coalition organized by state research and extension administrators, with help from lobbyist Ed Jaenke, under the auspices of the American Federation of Cooperatives, to seek increased funding for state experiment stations and extension services. In the second column is a 1978 coalition or-

TABLE 6.1: Groups Active in Agricultural Policy by Type

Type of Clientele Group	Coalitions				Testimony							Total	
	1979 Research		1978 Energy		Appropriations 1976		1979		Farm 1977				Total Groups	
	%		%		%		%		%				%	
General Farm Organizations	10	13	3	5	3	7	3	3	9	8			18	6
Commodity Groups	18	23	17	27	17	40	35	33	36	32			77	25
Trade Associations	13	17	36	58	11	26	16	15	28	25			82	27
Cooperatives	20	26	1	2	3	5	2	2	2	2			24	8
Corporations	5	6	4	6	0	0	1	1	18	16			26	9
Universities	0	0	0	0	0	0	15	14	7	6			22	7
Other Government and Program Groups	8	10	1	2	7	16	30	29	4	4			38	12
Professional Societies	4	5	0	0	2	5	3	3	2	2			11	4
Women's Farm Groups	0	0	0	0	0	0	0	0	6	5			6	2
Total	78	100%	62	100%	43	99%	105	100%	112	100%			304	100%

NOTE: Groups included are all those in clientele categories (contrasted with "public interest groups") which were active in joining one or both of two ad hoc coalitions, or which presented oral or written testimony to the House Agricultural Appropriations Subcommittee in 1976 or 1979, or which presented oral or written testimony before the House Agriculture Committee on the 1977 Food and Agriculture Bill. Some groups were active in more than one way, which explains why totals on the right may cumulate to less than the total number of groups identified with one of these activities.

ganized chiefly by a trade organization, the Fertilizer Institute, for the purpose of ensuring that the agriculture industry receive a high priority in federal allocation of energy. Also tallied are the private groups testifying before the House appropriations subcommittee on agriculture, in 1976 and 1979, and before the House Agriculture Committee on the Food and Agriculture Act of 1977. The number of groups testifying—which is rather large in each case—is but one indicator of group strength; indeed, many believe that farmers would be stronger if they could unite in a single farm organization.[1] In any case, however, agricultural interests find effective representation in a number of organizations.

General Farm Organizations

The general farm organizations, which invite membership from all types of farmers, have various orientations toward the research institutions. The National Grange was founded in the nineteenth century by public interest advocates whose objectives were similar to those of the founders of research institutions: to help farmers improve their farming practices and way of life. But the Grange soon became a protest organization without ever having developed much interest in agricultural research. Now a very small but visible group, the Grange supports agriculture research.

The National Farmers Union was founded early in this century by distressed farmers in the southern and Plains states. Farmers Union, with a wide range of liberal social concerns, has usually joined an electoral coalition in the Democratic Party. Its main objective has been to preserve the family farm, mainly through federal government support of high farm prices and through structural reforms, such as eliminating middleman monopolies. Two other groups of more recent origin have orientations similar to the Farmers Union's: the National Farmers Organization, organized in the late 1950s, and the American Agriculture Movement, organized in the late 1970s. All three groups have ambivalent views toward agricultural research: they recognize that it provides remedies against disease and blight and makes American products more competitive in international markets, but at the same time, they blame agricultural research for having produced the surpluses which hold prices down and therefore benefit consumers and agribusiness rather than farmers. (Most economists agree with these groups that farmers are not the major beneficiaries of research.) In most states, the public research establishment is seen as a handmaiden of the Farm Bureau. Thus, general farm organizations other than the Farm Bureau are likely to be suspicious and critical rather than supportive of research. The predominantly negative feeling toward research found within these groups was expressed by a staff member of the National Farmers Union: "Farmers don't need any more research." Except for a handful of Farmers Union state level organizations, the

Farmers Union, the National Farmers Organization, and the American Agriculture Movement were all absent from a recent coalition formed to support agricultural research.

The American Farm Bureau Federation, whose growth was intertwined with that of the state extension services (as described above), continues to be a supporter of agricultural production research, in part because of that special relationship with Extension, in part because it wishes to influence the direction of research, and in part because Farm Bureau's conservative leadership has considered research to be a more appropriate way for government to improve farm prosperity than price supports. Farm Bureau continues to support agricultural production research through its grass-roots committees, but the organization is no longer in a position to mobilize coalitions in behalf of research. Farm Bureau's conservative views on farm issues are generally incompatible with those of other general farm organizations, and the commodity groups in any case prefer to form their own ad hoc coalitions.

Commodity Groups

While some commodity groups include only the farmers or producers—for example, the National Association of Wheat Growers—most have a vertical organization (see table 6.2). Like the poultry groups, which include in their membership hatcherymen, feed companies, builders of chickenhouses, processors, and manufacturers of egg products, as well as a few producers, they tend to have industry-wide representation. In many groups, producers or farmers are no longer influential, both because their economic role in the industry is now relatively small and because processors or others are the more aggressive supporters of the commodity organization.

The vertically organized commodity group is based upon the assumption that all segments of the industry have common interests—for example, in preventing disease, in warding off burdensome government regulations, and especially in expanding production and markets. Theirs is a bullish posture that welcomes new technology, in contrast to

TABLE 6.2: Membership of Commodity and Trade Organizations[a]

Kinds of Members	Commodity		Trade	
	No. of Groups	%	No. of Groups	%
Farmers	23	42	5	7
(Open to farmers only)	(10)	(9)	(0)	(0)
Ag. Input Industry	5	9	13	19
Processors	10	18	15	22
Food Manufacturers	2	4	13	19
Distributors	11	20	11	16
Retailers	1	2	5	7
Scientists	2	4	3	4
Labor	0	0	0	0
Consumers / public	1	2	3	4
Total	55	101%	68	98%

[a] Includes those organizations active in one or both of two coalitions (1979 research coalition and 1978 energy coalition) or appearing at any of three hearings (1976 or 1979 House Agricultural Appropriations, or 1977 House farm bill), which were also listed in Nancy Yakes and Denis Akey, *Encyclopedia of Associations*, vol. 1, *National Organizations of the U.S.* (Detroit: Gale Research Company, 1979).

the old strategy of reducing production in order to increase prices and profits.

Commodity groups have an advantage in influencing public research in that, because research is organized according to commodity, their interests can be easily located in terms of specific projects, scientists, and research managers. The experiment stations tend to specialize in those commodities produced locally, as at Kansas State University, which has become a world center for wheat breeding, milling, and baking. And commodity groups are likely to possess the specialized knowledge upon which legislators depend. A survey of congressional staff members in 1979 (reported in table 6.3) indicated that 35 percent of these close observers viewed commodity organizations as "influential" upon research policy, while another 57 percent believed they were

TABLE 6.3: Congressional Staff Evaluation of Private Group Influence upon Agricultural Policy

	Percentage of Respondents Who Said:				
	Influential	Somewhat Influential	Not Influential	No Response	Total
General Farm Org.:					
General Ag. Policy	38%	58%	3%	1%	100%
Research Policy	9	74	13	4	100
Commodity Org.:					
General Ag. Policy	41	44	9	6	100
Research Policy	35	57	9	0	101
Ag. Trade Org.:					
General Ag. Policy	28	54	12	6	100
Research Policy	26	65	9	0	100
Ag. Business Corp.:					
General Policy	26	49	19	6	100
Research Policy	13	78	9		100

NOTE: Data is based upon a telephone survey conducted for *Successful Farming Magazine,* supervised by the author, administered by Paul Gardner in the summer of 1979 to 123 persons who were staff of the congressional agriculture committees or appropriations subcommittees for agriculture, or who served as agricultural specialists in the offices of individual members of these committees. Interviews were completed with 78 persons. The House appropriations subcommittee declined to permit its staff members to be interviewed. The number and proportion interviewed from other staff were as follows: Senate Agriculture Committee staff, 10 (83%); House Agriculture Committee staff, 8 (31%); Senate Appropriations Subcommittee staff, 2 (100%); assistants of senators on both Senate committees, 18 (58%); assistants of representatives on the House Agriculture Committee, 40 (77%).

Respondents were asked first to rate specific federal agencies and types of private organizations relative to their influence upon agriculture policy in general. Possible ratings were "influential," "somewhat influential," and "not influential." Respondents were then asked to choose one of five policy areas, including agricultural research policy, and to rate influence within that specific policy area. Twenty-three of the seventy-eight respondents chose to rate influence upon research policy in particular.

"somewhat influential." Only 9 percent thought they were "not influential." In contrast, the number who viewed general farm organizations as not influential was larger than the number who viewed them as influential.

Industry Groups

Trade Organizations

There are no clear lines distinguishing the commodity groups from the trade associations, which speak for providers of particular services within the agricultural industry. For example, the National Canners Association, a trade association, has been instrumental since 1926 in organizing agricultural research to develop vegetable varieties for canning. Trade groups may be "progressive" in this way, but they may be quite "unprogressive" in the face of potential technological change that would reduce the value of their own output. Thus, the National Agricultural Chemicals Association deems "unrealistic" those pest management strategies which would sharply reduce the use of pesticides.

Both trade associations and commodity groups want public research institutions to do research from which their members can generate a successful product. They are therefore likely to have very specific research interests, and may employ scientists qualified to advocate those interests. In explaining this, one trade association official drew a distinction between his staff and those of the general farm organizations: "They hire the wrong people, people who can only come and say 'we have 300,000 farmers.' We hire young men, some with Ph.D.'s, who can state the costs and benefits of our product, and that argument carries the day in Congress now." Trade and commodity groups frequently express resentment that their scientific talent and experience are not more routinely tapped in research decision making. But as we shall see, these groups do generally have good access to the research institutions.

Agricultural Corporations

Agribusiness corporations range in size from relatively small family operations to the very large organizations which are often accused of dominating markets.[2] In the several sectors of food processing and merchandising, the four largest firms collectively averaged 39 percent of total shipments, as of 1970.[3] Because these firms are so large, some industry critics insist that they should not receive assistance from public research. Much research is done by the corporations themselves, even

96

as they seek to direct public research in their own interest. By a conservative estimate, private agricultural research constitutes 40 percent of total agricultural research expenditures: $808 million (1976), compared with $636 million (1979) by the USDA, plus $156 million (1976) in federal grants to the experiment stations, and $385 million (1975) in state expenditures.[4] However, much of this private research is for product development. A major hybrid seed company's research budget, for example, is spent mainly in adapting corn seeds to local environments, while the company depends on public sector research for such tasks as breeding for insect, disease, and drought resistance, developing minor seeds for minor markets, and, of course, basic research in plant genetics. A large chemical company, according to its spokesman, devotes major effort to "looking for compounds that have some value as pesticides." Representatives of private research stress that their management "must see a return on research investment. That's the overall climate in which we are operating." It follows that most private research is for "applied" and "developmental" research,[5] the results of which the companies do not intend to share with the public research community.[6] The occasional investment in long-term research "is so easy to cut off when the industry has a financial crunch—and it frequently does," said one private industry research leader. "If you don't survive tomorrow you can't do any research." Therefore agribusiness is heavily dependent upon public research. Many firms provide modest grants to public agencies to induce them to commit resources for the "basic" research from which these firms may develop their products.

It appears that industrial firms spend relatively little of their own funds on environmental or food safety research, or for human nutrition research.[7] Rather, they have increasingly committed their funds to "defensive" research designed to protect them against new regulations dealing with the adverse side effects of agricultural production techniques.

Private industries, trade associations, and commodity groups, too, are understandably eager to assure that public research institutions

97

take a "defensive" posture rather than one supportive of unfavorable regulations or education. An official of a state cattlemen's association exemplifies this point of view: "We have been frustrated by the different voices on nutrition at the University. The Home Economics Department is sympathetic to the McGovern Committee's dietary goals [which specify less beef in diets], and their viewpoint tends to get into the hands of Extension and drifts into 4-H. But there are different views in the College of Agriculture, and I am pleased to know there is to be a discussion of University philosophy soon."

Cooperatives

Farm cooperatives, too, are agribusiness firms. Because they are becoming the major providers of synthetic fertilizer, they are interested in research which maximizes its usefulness. Cooperatives have limited but specific research interests, and they expertly represent those interests.

Agency Groups

A number of government agencies—local, state, and federal—are clients of agricultural research, and these agencies themselves may have clientele organizations. The National Association of Soil Conservation Districts (NASCD), for example, represents special district governments across rural America which were formed in response to a federal soil conservation law: "We are one of the few organizations that crosses that boundary between the agricultural establishment and the environmentalists," notes NASCD's president.[8] Similarly, the National Association of State Departments of Agriculture represents a coalition of research users.

Many of the "user" government agencies do not have national associations which speak for them. Among these, one of the most important "clients" of agricultural research is the federal Animal and Plant Health Inspection Service (APHIS). Federal research agencies seek to develop support from these user agencies within the government.

The research institutions also lobby in their own behalf. A well-established national organization, the National Association of State

Universities and Land Grant Colleges (NASULGC) speaks for the large public universities. There are two semiautonomous committees of NASULGC, the Experiment Station Committee on Policy (ESCOP) and the Extension Committee on Policy (ECOP), which speak, respectively, for the experiment stations and the extension services. There is a tension between the parent NASULGC, in which the university presidents serve the interests of both nonagricultural and agricultural scientists, and the two committees, composed of agricultural research administrators, which champion production-oriented research. The tension erupted in 1979 when the committees broke away to organize a lobby for their own budget.

WHERE AGRICULTURAL GROUPS HAVE INFLUENCE

When officials in a number of agricultural groups were asked in 1978 to name those public research institutions which were doing research of interest to them, most of them named one or more state experiment stations, and some named the federal Agricultural Research Service (ARS) or Economic Research Service (ERS) (ERS briefly became the Economics Statistics Service). Other institutions mentioned were the state colleges of veterinary medicine, the Tennessee Valley Authority (for fertilizer research), and the Rockefeller and Ford foundations.

When asked how well their expectations had been fulfilled by the state experiment stations, the ARS, and the ERS, officials of trade groups were likely to rank each of these institutions as "high," while commodity groups were likely to rank them as just "acceptable." Yet there was apparent agreement among organization spokesmen in both categories that it was the commodity groups which had better access to the research agencies. An explanation for this is that commodity groups had higher expectations. As one commodity group leader said, "Let's don't kid ourselves. Our industry is a shining example of the success of agricultural research." Yet this leader ranked the research agencies only as "acceptable." He believed that ERS was too slow, that the experiment stations tended to go overboard for the major state commodities (too often not including his own commodity), and that public funding for research on his commodity was generally not commensurate with its high value in the marketplace.

Group representatives were shown a list of decision makers, and were asked whom they contacted to make their research needs known. The list of decision makers included the president, the Office of Management and Budget, state governors, college presidents, the secretary of agriculture, the assistant secretary of agriculture for research, heads of research agencies, college deans, academic department heads, research team leaders, individual researchers, and committees of state and national legislative bodies. Although there was some variety in responses, several patterns of contact did emerge. At both state and national levels, commodity and trade groups have organized their own research agenda committees, on which one or more public research scientists usually serve. The scientist member helps decide research needs, helps communicate these needs to the institutions, and, in the case of smaller state groups, may be the researcher who actually does the work. In any case, there is likely to be much informal contact with the researchers themselves.

In the case of state-level commodity groups which represent one of the state's major products, the governing board may include the chairman of the relevant academic department. Said one official of a state commodity organization: "The chairman of the department is on our board and works closely with us. He can go to the dean for us, or can write up proposals as to what we want to do and get his department working on it."

At both state and national levels, groups representing a major commodity may also consult with one of the principal administrators. In a federal research agency, this might be the area or regional director, the national administrator or one of his assistants, or even the assistant secretary; at the state level, the list would include the agriculture dean or experiment station director (sometimes one person holds both of these offices). While some groups begin at the lower decision-making level (the scientist) and move upward as necessary, others begin at the top. Said one of the latter: "We talk to some subject-matter people, but actually research people have their hands tied as to the amount of emphasis that people above them desire. So it is more effective to deal at the administrative level—the dean and director—rather than at the specialist level."

For virtually all commodity and trade groups, frequent and more-or-less friendly contact with the researchers or the research administrators—or more likely with both—were typical strategies for influencing decisions. In a study of Kansas experiment station scientists, 40 percent of the agricultural researchers said that they "had experienced pressure to research particular topics," while 30 percent had felt pressure *not* to research particular topics. Of the sources of pressure, half were from inside the institution—experiment station directors and department heads—with the other half from outside organizations or individuals.[9]

Another common strategy described by national group officials was to contact a friendly senior legislator who could intercede both within the administration and to assure funds in Congress. Each national commodity group seemed to have one or more champions within Congress who would fight for its appropriations. And national groups routinely expressed their needs in testimony before the agricultural appropriations subcommittees.

The major commodity and trade organizations were also ready to go, "when we need to," to the White House, to OMB, and to the chairman of the House Appropriations Subcommittee, Jamie Whitten (D., Miss.). "Yes, we do talk with OMB and the White House Staff, and directly with the president," said one spokesman for a major commodity group. "Our board has met several times with President Carter and with his foreign trade representative." The directors of two trade associations each stated that they were personal friends of Congressman Jamie Whitten's, and both said they might discuss business with him during a golf game, whenever necessary. Whitten and the subcommittee's former minority leader, Mark Andrews (R., N.Dak.), were the only members of this important subcommittee to whom the groups who were surveyed had talked. However, several group leaders stressed that research issues were seldom among the high priority issues to be discussed with the president or with a committee chairman.

On issues important to a number of agricultural groups, coalitions have been formed. In the past, such coalitions were organized by general farm organizations. Today they are organized on a single-issue basis by the groups which are most affected. For example, an "energy users" coalition was organized during the energy crisis by the Fertilizer Institute and others for the purpose—eventually achieved—of persuading the administration to give highest priority to agricultural uses of oil. Private groups were also mobilized, successfully, in behalf of federal funding for state research and extension, as described below.

MOBILIZING PRIVATE GROUPS
IN BEHALF OF STATE RESEARCH

In 1978 and 1979, a number of state extension and experiment directors brought about an effective coalition of "aggies" (agricultural / industry groups) in support of research and extension budgets. In the minds of these research heads, a crisis had arisen. Federal grants-in-aid for state research had been declining, both in terms of constant dollars and also by comparison with federal support of various nonagricultural research missions. The Carter administration, following the advice of various research evaluating committees, preferred to offer any increased funding through an expanded competitive grants program, rather than through institutional budgets, with the competitive grants to be awarded by scientist peer groups, rather than by research administrators. Further, institutional budgets, which had traditionally been unregulated, were being shaped to reflect national priorities.

The state research administrators decided to hire a lobbyist, but their parent organization, NASULGC, refused to serve as a medium for direct lobbying. Despite opposition from NASULGC, the agricultural committees within the organizations (ECOP and ESCOP) raised $100,000 on their own and employed a prominent agricultural lobbying firm, Ed Jaenke and Associates.

102

Jaenke's strategy was first to mobilize the latent political strength of the experiment stations and extension services, whose administrators had been divided by competition and factionalism. These administrators were urged to join forces in support of an omnibus federal budget for research, extension, and teaching in the state colleges of agriculture. Jaenke's second step was to generate support from other agricultural groups. He argued to these "aggies" that more research was needed to bolster agriculture's slipping growth rate.

When the coalition was formed at year's end in 1978, the administration budget had been readied for submission to Congress, so the coalition began lobbying in Congress, at first in the agriculture committees. They began with the agricultural committees because, under new congressional budgeting procedures, the standing committees are expected to indicate the budget needs in their jurisdictions. In this case the agriculture committees did recommend increased research funding. Taking note of these recommendations, the budget committee in each house then made its recommendations, and the Senate's budget committee included a statement supporting increased agricultural research. With this "snowballing support," a successful appeal was then made to the Senate and House appropriations committees. Whereas the administration's 1980 budget for state agricultural research and extension had proposed to hold spending at the previous year's level, the congressional appropriations process increased this amount by 14 percent. Although this increase was accomplished simply by shifting funds from the competitive grants budget to the institutional budgets, the result was hailed as a victory for the research coalition. Most state administrators felt that an even larger increase was justified, in order to compensate them for "losses" in earlier years. It was Jaenke's hope that the administration, having observed that its budget recommendations had been overridden, would be reluctant to try to reduce funding for state institutions in future years.

This victory had been gained, however, by returning to the narrow political base of support from agricultural groups. By sacrificing the competitive grants desired by nonagricultural groups, the "aggies" had

103

ignored a larger coalition which had been fashioned by NASULGC and others in the passage of a new research charter in the 1977 Food and Agriculture Act. This broader coalition (discussed in chapter 8) included, among others, environmentalist and consumer groups.

This "aggie" victory may rest on an inadequate base. In a study of subsystem support for research agencies within the House Appropriations Subcommittee for Agriculture, using the number and intensity of witnesses before the subcommittee as a measure, political scientist Ken Meier found that both the federal Agricultural Research Service and the Cooperative States Research Service had been more successful than other USDA agencies in generating support before that subcommittee (see table 6.4). Support for research had come mainly from agricultural groups. But Meier found that support for all USDA agencies had decreased during the past two decades.

Meier noted that although the ARS had generated greater clientele support than had other USDA agencies, the agency had not experienced relatively greater growth. He concluded: "If stability is the ARS's reward for clientele support, it is a very modest reward."[10]

Meier, in comparing the budgetary success of the various agencies with the amount of clientele support they received, could not find evidence that support from the subsystem clientele had been helpful. In the last analysis, he questioned "whether the bureaus benefit from strong clientele support." Some scholars believe, on the one hand, that bureaus with strong clientele support receive additional money, and autonomy in the use of that money. But Meier found, on the other hand, that interest group support "is not an unmitigated blessing." In fact, he noted, "students of regulatory policy in particular feel that interest group support may well be harmful to an agency."[11] Observing that regulatory agencies have often turned to their clientele after having lost the broader or "diffuse" support which had led to their creation, Meier suggested that such a strategy frequently proved counterproductive: "Since strong clientele relationships with the regulated [groups] are perceived as illegitimate by the major policy actors, Congress and the President respond to the captive agency by ignoring

TABLE 6.4: Clientele Support of USDA Agencies before House Appropriations Committee
1974–1976

Agency	No. of Groups	Intensity[a]
Research Agencies:		
Agricultural Research Service	72	2.53
Economic Research Service	3	2.67
Statistical Reporting Service	16	1.75
Cooperative States Research Service	41	3.12
Extension	14	3.14
Other USDA Agencies:		
Animal and Plant Health and Inspection Service	26	2.65
National Agricultural Library	0	0
Commodity Futures Trading Commission	0	0
Packers and Stockyards Administration	7	1.29
Farmer Cooperative Service	1	3.00
Foreign Agricultural Service	16	3.13
Export	0	0
Agricultural Stabilization and Conservation Service	27	2.11
Federal Crop Insurance Corporation	1	3.00
Commodity Credit Corporation	2	4.00
Rural Development Service	3	3.00
Rural Electrification Administration	10	3.60
Farmers Home Administration	25	2.72
Soil Conservation Service	42	2.40
Agricultural Marketing Service	12	1.42
Food and Nutrition Service	15	.60

SOURCE: Kenneth John Meier, "Building Bureaucratic Coalitions: Client Representation in USDA Bureaus," in Don F. Hadwiger and William P. Browne, eds., *The New Politics of Food* (Lexington, Mass.: Lexington Books, 1978), table 6–2.

[a] Testimony supporting the agency in general was coded +3; that supporting an agency program was coded +2. A group which supported both the agency and its program was therefore scored +5, while one which testified negatively on both was scored −5.

its needs, cutting its budget, and constantly meddling in its administrative activities."[12]

Indeed, agricultural research institutions had already experienced such a challenge to their legitimacy because of their dependence on groups within the agricultural subsystem. NASULGC, in developing the broad coalition that passed the 1977 agricultural act, had participated in an effort to restore legitimacy by reconciling agricultural production goals with other goals, such as environmental protection, nutrition, and food safety. Agricultural administrators had abandoned this conciliatory posture for a return to subsystem politics.

Industry groups will continue to be influential in research decision making, in part because they have achieved increased funding for research, and in part for other reasons. As noted in chapter 4, researchers want to do something useful, and satisfied clients assure them that they are succeeding. Still another source of influence for agricultural groups and corporations is the "resources," in the form of status and money, they are able to contribute. Trade and commodity groups sometimes underwrite prizes given by professional societies for designated kinds of research, and a few private corporations give large cash awards for useful research. Agribusiness corporations also commonly employ as consultants researchers who have done work interesting to the corporation, while land-grant administrators frequently serve as paid directors of major agribusiness corporations.

Industry research funds are granted directly in several forms. Major companies within an industry may make "foundation" grants to those universities which produce the specialists and specialized research on which the industry relies, as well as thousands of small grants for specific research. Also, there are now more than a hundred commodity taxes, levied at the initiative of the particular industries, which are allocated by an industry-dominated committee for specific research and promotion activities. In California, commodity groups have preferred to raise millions of research dollars through these taxes rather than to battle with public advocacy groups for a larger share of general revenue units. Fujimoto and Kopper conclude that they "have abandoned legislative lobbying and adopted marketing orders as a means of insuring that needed production research is completed by the University of California."[13] Because the decision process for research which is funded by these commodity taxes is likely to be open and explicit, through its use of the peer review format, such funding more clearly reflects an industry-wide perspective than do the direct grants that encourage the interests of particular firms.

The pervasive presence of private influence within the state agricultural research institutions was the theme of Jim Hightower's exposé, *Hard Tomatoes, Hard Times,* published in 1972. Hightower believed that this presence helped explain why agricultural colleges had so readily abandoned their function of serving family farmers in favor of providing services for the large agribusinesses. He also argued that this was why they had ignored the human and environmental costs of the "violent" technological revolution they had wrought in the countryside. Hightower concluded: "The land grant colleges must get out of the corporate board rooms; they must get the corporate interests out of their labs."[14]

Hightower's book stirred anger within the establishment, and it created public interest in the problem. In the ensuing years, research decision making was opened to a wider public, including critics other than Hightower, and research institutions took pains to associate their research with "societal" missions.[15] Yet private funding continues at a high level, and is increasing with the spread of the commodity tax. When general appropriations for research do not grow, funding from private corporations or from earmarked taxes is seen as a way to fill the gap.

THE MASONEX PROJECT

The Masonex Project is one example of private funding of public research. Its sponsor, the Masonite Company, was founded on a process developed by William Mason in the 1920s which utilized a waste product—sawdust—to make wallboard. There was a waste product, too, from the production of wallboard: a water soluble glucose extracted from the sawdust, for which the Masonite Company wished to find a commercial market. After various efforts to find uses for it had failed, it was developed in 1959 as a cattle feed. This product, called Masonite, made Masonite an agribusiness firm.

But this waste product, still accumulating as a mountain of ooze in a northern California forest, was seen to have still other potentialities. Masonex contained tannin, once used to tan leather, and tannin might prove useful in preserving proteins consumed by a cow until they could be properly digested in the cow's stomach. There was also a prospect that this waste product might contain a "growth agent," one of those substances emitted by the body's glands which instruct the body to grow.

Masonite pursued these exciting leads in the laboratories of several land-grant scientists. For the most part, the scientists asked to cooperate were those whose work in the area had come to the attention of a Masonite executive, Dale Galloway. Galloway offered each scientist a grant of between $4,000 and $10,000; typically, these funds from Masonite were merged with much larger amounts of state and other public funds, in research programs with multiple objectives. For example, in one experiment using a $4,000 Masonite grant, there were more than two hundred cattle, which required $150,000 a year to maintain. Only forty-nine head of the cattle were used in the Masonite experiment. In this case, the professor had contacted Galloway to seek use of his material. He had also asked for a small grant to cover the stipend of a research assistant. Galloway had been fully responsive.

Why seek funding from Masonite? Recipients gave several reasons. Several were seeking a way to apply their own research findings. Said one scientist: "An idea, once tested, must be put into motion, usually to produce some kind of product that is handy for the farmer to use." Masonex, as a byproduct, would be a relatively inexpensive resource for producers. And if it did trigger growth in meat animals, it could contribute to world food supplies. Working with a private firm enabled graduate students to be in touch with the world outside the university, in which many would be seeking jobs. For a few scientists private funding was a necessity: all financial support for graduate students came from small private grants because no public funds were available for that purpose. Several scientists explained that money for training and research programs has to come from somewhere.

Indeed, Masonite money was preferable to public funds, in that it was relatively unrestricted. For one professor, Masonite money had made it possible to do a project which the federal Food and Drug Administration had once approved but had lacked funds to finance. Another professor was able to carry out a project which the National Science Foundation had refused to grant, "because the peer reviewers didn't understand growth factors." Therefore, commented this professor, "we appreciate the Masonite money very much, indeed more than I have indicated to you." Further, Masonite assembled its researchers together each year, "usually in a very pleasant location such as San Diego," where there was a useful rubbing of elbows with people from different industries.

All the recipients of Masonex funding noted the danger of cooperating with private business. But they were inclined to differentiate their own relationship from that, for example, of scientists who profited from using the university's name and resources in uncreative research performed primarily for the benefit of the industry.

Some Masonex cooperators reported that their experiment stations had no ethical codes to guide researchers: "Everything is negotiable." However, others said their institutions had listed conditions for accepting research such as the following:

1. The intended use of private funds should fit into the overall mission of the department or college.
2. The grant must never be between an individual and the corporation or industry, but instead between the research institution and the industry.
3. The researcher rather than the industry must determine the research design.
4. The institution must be the patenting entity in any new development.
5. The information developed should be made available by the pro-

fessor to the department and the college and to relevant areas of the public, including even any adverse findings on a company's product.

6. The university's name may not be used as part of an advertising program.

Such rules, however, do not cover all cases, and may suffer in application. Several professors said that the best hope for achieving a proper relationship is to rely on the integrity and judgment of the professor, while recognizing that "if you have dingdongs who are willing to take anything that comes along, you will get problems."

Beyond the question of whether specific cases are appropriate, there is a cumulative effect upon institutions from cooperation with private industry: the notion that there is a "public interest" can be lost. Said one research leader involved in the Masonex project: "We know there is a tendency of the industries to try to dominate university research in terms of the kinds of use that will be done, and also then to use the university to agree with their own preferences about the product." And on the university's side he noted: "The heads of research are conservative, they are oriented toward the private sector, they are never critical of what it is doing. University professors, too, are usually blindly on the side of industry, for example, as against criticisms from the Environmental Protection Agency. There is a lack of objectivity on the part of the professors and the college."

So the influence of private groups through direct support must be evaluated at two levels. At the level of the individual grant, each of the scientists who received Masonex grants perceived them as ethical and indeed beneficial to the research institution and to the public interest, even as they were judging that some other private grants, perhaps those given to colleagues, were not in the public interest. At the institutional level, and even at the level of academic disciplines, there has been a failure to make distinctions between the interests of public research and those of private organizations.

An example of institutional diversion through private grants, unrelated to Masonex, is the pursuit of hybrid seeds, encouraged by seed company grants to scientists and to their academic periodicals. Hybrids serve the private sector's interest because companies are able to keep control of parental lines and profit from them in successive years. Public decision makers over the years have made an easy presumption, lately challenged, that there are no serious disadvantages connected with hybrids.[16]

A painstaking study of the University of Nebraska's Department of Animal Science has provided other evidence of the cumulative influence of small private grants.[17] This study, by the Center for Rural Affairs, revealed that most private grants to the department were for applied research in animal nutrition and for product testing or product or market development in the field of animal nutrition (see table 6.5). These grants were an important increment covering research assistantships and other vital needs. This "demand" from private firms, bolstered by financial incentives, appeared to be a reason why animal nutrition had come to be favored over other specializations within the department in the allocation of staff and other research support.

In this and other cases, the issue is not mainly whether private support has a good or deleterious effect upon the assisted research; indeed, there is apparent complementarity between the excellent record of Nebraska's animal nutrition research and the high interest that numerous private firms have displayed in it. The main issue, rather, is whether research decisions should be heavily influenced by profit seeking, when, in a certain theoretical sense, producers and consumers would be best served by a prolific agriculture which buys virtually nothing. In this age of limited resources, distinguished research panels as well as public interest groups seem to be urging researchers to seek input-reducing "miracles," such as nitrogen fixation and enhanced photosynthesis, which, from the viewpoint of many agribusiness industries, appear to hold more threat than promise.

COMMENTARY

The cumulative representation of private groups within public re-

111

TABLE 6.5: Summary of Private Grants by Research Area, 1981 Dept. of Animal Science, University of Nebraska-Lincoln

Research Area	No. Projects	Product Testing	Product or Market Development	Professional or Career Development	Management Advice or Consumer Report	General Support	Total
			Amount of Grant by Purpose				
Nutrition:							
Beef	4	94,287	192,972	0	0	13,847	301,106
Dairy	1	10,000	7,000	0	0	0	17,000
Poultry	3	31,914	6,000	0	0	1,000	38,914
Swine	3	26,020	95,160	0	13,250	0	134,430
Subtotal	11	162,221	301,132	0	13,250	14,847	491,450
Breeding:							
Beef	2	0	0	1,600	300	2,100	4,000
Dairy	2	0	0	0	0	615	615
Swine	1	0	0	0	0	0	0
Subtotal	5	0	0	1,600	300	2,715	4,615
Physiology:							
Beef	1	0	0	0	0	0	0
Dairy	3	0	0	0	15,550	0	15,550
Poultry	1	0	0	0	0	0	0
Swine	3	0	0	0	16,500	0	16,500
Subtotal	8	0	0	0	32,050	0	32,050
Meats & Products:	3	0	181,254	3,540	0	0	184,794
Subtotal	3	0	181,254	3,540	0	0	184,794
Miscellaneous		0	12,500	2,595	0	15,810	30,905
Total	27	162,221	494,886	7,735	45,600	33,372	743,814

SOURCE: Center for Rural Affairs, Walthill, Nebraska.

search agencies has been pervasive. One corporation spokesman named seven institutionalized means for regular industry input into research administration: on-campus visits, local task forces, national agency reviews, accreditation reviews, representation on the Joint Council of the Science and Education Administration, industry-university personnel exchanges, and industry sponsorship of research. In addition, there exists the Agricultural Research Institute, organized in 1951 as "a crossroad where agricultural research managers in government, industries, and universities meet and mingle for full, free, and wide-ranging discussion of the nation's agricultural research programs and needs."[18]

Such industry representation is needed because, in our system, public research and private production are interdependent. As we have seen, big agribusiness does not do much of its own research other than for product development, nor will it begin to do so just because others think it ought to. The technological progress which has gained the agricultural industry its reputation for efficiency is to be credited largely to public research institutions. Moreover, there is little reason to assume that agricultural and business interests within agriculture intended, or, on the whole, profited from, the overall pattern of technological innovation in agriculture. The major political activity of farm groups has been to seek higher prices through regulation which would control agricultural production. Indeed, prices are expected to be generally higher in regulated markets, as one study indicates for medical care,[19] but research helped prevent that effect in American agriculture. For agribusiness industries, too, public research may not on balance be a means to higher inputs and higher profits. It has begun to look as if the dramatic impacts of public agricultural research are not controllable by groups within the industry.

In other respects the interests of private business and public research are not identical. The private corporation's interest in making a profit is not always complementary to such state and national legislative goals as preserving the family farm, protecting the safety and health of agricultural workers, guaranteeing food safety, dispensing nutrition edu-

113

cation, economizing in the use of energy, revitalizing rural communities, providing equality to rural minorities, and assuring that all citizens have adequate food.

Many of these social objectives have been ineffectively represented within the subsystem. For the most part, the federal agencies charged with protecting citizen health and safety are outside the USDA; they have been ignored or held at arm's length from the process of determining agricultural research policy. Environmentalists and other public interest groups have definitely not been invited to sit at the conference table with scientists, administrators, and representatives of the agricultural industry.

In recent years, groups representing interests other than those of commercial agriculture firms have occasionally been granted formal means for representation. Consumer and environmental representatives were first invited to a USDA-sponsored conference on research alternatives held in Kansas City in 1976. The 1977 federal research law created an advisory council of "users" which included representation for consumers, environmentalists, and farm workers. The Carter administration named, as assistant secretaries of agriculture, several persons who previously had been spokespersons for consumers, environmentalists, and the rural poor. But within the ranks of institutional leadership there remains the problem of representing interests beyond those of industry, and the same problem is found in the congressional process for public funding of agricultural research.

CHAPTER SEVEN

The Congressional Appropriations Subcommittees for Agriculture

Congress has worked its will on research policy primarily through the power of the purse. Until 1977 research law was permissive and open-ended, so that research was primarily guided by the dollars allocated by the congressional appropriations subcommittees in their oversight of agricultural research. The predominant influence of these appropriations subcommittees is unique to agricultural research policy (as opposed to other aspects of agricultural policy), according to evaluations by congressional staff members presented in table 7.1. As the table shows, 91 percent of congressional staff surveyed agree that the House Appropriations Subcommittee is influential over research policy, and 83 percent think that the Senate Appropriations Committee is influential.

Other congressional committees have also influenced research policy, and some of those which reflect the interests of a nonindustry coalition will be noted in the next chapter. The congressional agricultural committees, traditionally part of the agricultural subsystem triangle, have begun to be influential as agencies integrating a grand coalition that embraces both industry and nonindustry interests, as evidenced in the passage of the 1977 research law. But it is the appropriations subcommittees for agriculture, rather than the ag-

TABLE 7.1: Percentage of Congressional Staff Rating Committees as "Influential" Over Policy Areas

	No. of Respondents[a]	House		Senate	
		Ag. Comm.	Approp. Subcomm.	Ag. Comm.	Approp. Subcomm.
Ag. Policy in General	78	81	77	81	67
Ag. Research	23	65	91	57	83
Grain Policy	20	75	35	75	35
Combined Specific Policy Areas	73	70	62	67	59

[a] Based on a telephone survey conducted for *Successful Farming Magazine,* supervised by the author, administered by Paul Gardner in the summer of 1979 to 123 persons who were staff of the above committees or who served as agricultural specialists in the offices of individual members of these committees. Interviews were completed with 78 persons. The House Appropriations Subcommittee declined to permit its staff members to be interviewed. The number and proportion interviewed from other staff were as follows: Senate Agriculture Committee staff, 10 (83%); House Agriculture Committee staff, 8 (31%); Senate Appropriations Subcommittee staff, 2 (100%); assistants of senators on both Senate committees, 18 (58%); assistants of representatives on the House Agriculture Committee, 40 (77%).

Respondents were asked first to rate specific committees relative to their influence upon agricultural policy in general. Possible ratings were "influential," "somewhat influential," and "not influential." Respondents were then asked to choose one of five policy areas, including agricultural research policy, and to rate influence within that specific policy area. Twenty-three of the seventy-eight respondents chose to rate influence upon research policy.

riculture committees, which mainly wield the agricultural subsystem's power within the Congress.

THE SENATE APPROPRIATIONS SUBCOMMITTEE ON AGRICULTURE

Table 7.2 lists the membership of the Senate appropriations subcommittee since 1958. Though many states have been represented through these years, a few members have been much more influential than others, either through some combination of long service and service as chairman or senior member of the minority party, or because of extraordinary commitment to the subcommittee.

116

TABLE 7.2: Senate Appropriations Subcommittee for Agriculture: Membership for 1958 through 1981

Previous Members (1958–80)

Democrats	Year Began	Years Served		Republicans	Year Began	Years Served
Russell (GA)	1958	13		Young (ND)	1947	32
Haydn (AZ)	1958	11		Dirksen (IL)	1953	6
Hill (AL)	1958	11		Ives (NY)	1958	1
Robertson (VA)	1958	9		Hruska (NB)	1959	10
McGee (WY)	1959	17		Schoeppel (KS)	1961	2
Dodd (CT)	1959	2		Case (NJ)	1963	4
Johnson (TX)	1959	2		Javits (NY)	1967	2
Humphrey (MN)	1961	2		Fong (HI)	1969	8
Mansfield (MT)	1963	3		Boggs (DE)	1969	6
Proxmire (WI)	1965	14		Hruska (NB)	1971	6
Yarborough (TX)	1965	5		Mundt (SD)	1971	4
Inouye (HI)	1971	8		Hatfield (OR)	1973	6
Mansfield (MT)	1971	2		Bellmon (OK)	1973	6
Hollings (SC)	1973	2		Stevens (AK)	1977	2
Bayh (IN)	1975	4		Garn (UT)	1979	2
				Schmitt (NM)	1979	2

1981 Membership

Democrats	Year Began	Years Served		Republicans	Year Began	Years Served
Eagleton (MO)	1973	9		McClure (ID)	1979	3
Stennis (MI)	1955	27		Andrews (ND)	1981	1
Byrd (WV)	1969	13		Abdnor (SD)	1981	1
Chiles (FL)	1975	7		Kasten (WI)	1981	1
Burdick (ND)	1977	5		Kattingly (GA)	1981	1
Sasser (TE)	1979	3		Specter (PA)	1981	1

For many senators, the subcommittee assignment has not been a matter of high personal priority, as can be inferred from the fact that its membership has included successive Senate majority leaders and many other illustrious members who were preoccupied with other roles. In 1959, for example, even the subcommittee's chairman, Sen. Richard Russell (D., Ga.), had other powerful roles as the South's

117

TABLE 7.3: Senate Additions to the 1978 Research Budget:
Three Low-Knowledge Strategies for Exercise of Power

1. Member Expertise

Research Project	Interested Expert Member
ARS:	
Contingency fund	Bellmon, Okla.
Tanning research	Bellmon, Okla.
Increase ARS contingency fund	Bellmon, Okla.
Brucellosis	Bellmon, Okla.
CSRS:	
Range caterpillar[a]	Bellmon, Okla.

2. Home State Interest

	Interested Member	Basis of Interest[b]
ARS:		
Soybean research increase	Eagleton, Mo.	Impact
Dairy Forage Research Center	Proxmire, Wis.	Location
Human Nutrition Laboratory increase	Young, N.D.	Location
Soil Erosion Laboratory	Bayh, Ind.	Location
El Reno Research Center	Bellmon, Okla.	Location
Plant stress and moisture	(Bentsen, Tex.)[c]	Location
Onions, cucumbers, carrots	Proxmire, Wis.	Location
Tanning research	Bellmon, Okla.	Impact
Pseudo-rabies	Bayh, Ind.	Location
Oil and Water Research Lab	Byrd, W. Va.	Location
Lone star tick	Bellmon, Okla.	Location
Northern Great Plains Research Station	Young, N.D.	Location
Mint research	Hatfield, Oreg.	Impact
CSRS:		
STEEP (Solutions to Environmental and Economic Problems—Oreg., Wash.)	"Several committee members"	Impact
Range Caterpillar	Bellmon, Okla.	Impact

regional leader in the Senate and as chairman of the Armed Services Committee. In 1978 Sen. Thomas Eagleton became subcommittee chairman by virtue of the fact that a majority of the subcommittee's Democrats, all senior to Eagleton, had chosen more prestigious leadership roles in the Senate.

118

TABLE 7.3, continued

---

3. *Outside Authority (New Agenda Programs)*

New Agenda Programs

ARS:
Competitive grants for basic research
Nutrition Research Lab, Tufts Univ.
Nutrition Research Lab, Baylor Univ.
Nutritional Impact Research

CSRS:
STEEP[a]

Extension:
Title V Rural Development Research
"Direct" marketing of farm produce

Other Senate Additions, Not Classifiable Above

Restore House cut in general ARS budget
Restore House cut in GSA space rental charges
ARS maintenance of research facilities
Increase CSRS above administration and House recommendations
Increase Extension above administration and House recommendations
Citrus blackfly
Stem rust in wheat
Fever tick

SOURCE: Senate Report no. 95–296 (95th Cong., 1st sess.) on Agriculture and Related Agencies Appropriation Bill, 1978. Explanations were gained from the report, previous hearings, conference, and supplementary interviews.

[a] Cited also under category 2.

[b] In most cases, the home-state would benefit from both the location and the impact of the research; the category listed is the one which was emphasized by the Senate.

[c] Not a member of the subcommittee.

Kenneth Meier has pointed out, in an understatement, that the Senate appropriations subcommittee "is less specific in its annual examination" of agricultural research than is the House subcommittee.[1] In one three-hour session during the 1977 Senate hearings, for example, the Senate subcommittee heard the testimony of both major research agencies—the Agricultural Research Service and the Cooperative States Research Service—plus that of the Rural Electrification Administration. In the House appropriations subcommittee, by contrast, an entire morning was devoted to the Agricultural Research Service. Attendance at Senate hearings is low: at one point in the 1977 hearings, Chairman Eagleton exulted that five of the twelve members were present—"a high-water mark for this committee."[2] Even those members who were seeking research facilities for their states often chose to submit written requests rather than to appear personally at the hearings.

Thus the Senate subcommittee is an anomaly of power lacking infomation. Table 7.3 suggests three low-knowledge strategies used by the subcommittee in 1977 to exert influence upon the research budget. One strategy was to support the occasional "expert" member, in this case a junior member of the minority, Sen. Henry Bellmon (R., Okla.), who was able to argue knowledgeably in favor of a number of budget changes.

For other members, the more typical strategy was to "bone up" on one or more projects important to their own states, and to argue convincingly for these. Support for this "pork barrel" orientation was evident in the sponsorship of several amendments listed in table 7.3. This orientation predominated at the 1977 "mark-up" session, in which subcommittee members ironed out the report on the appropriations bill. At that session members participated by seeking commitments of funds for research facilities to be located in their home states or for research projects which would be of service to commodities produced in their states. Results of this pork barrel approach over a longer period are evidenced in table 7.4, which shows that 44 percent of all research construction authorized between 1958 and 1977 was to be done in states of sitting members of the Senate subcommittee. From this table we may also note that members of the House subcommittee were not similarly favored, except for the House chairman, Jamie Whitten, whose district re-

TABLE 7.4: Placement of Agricultural Research Service Construction Projects—1958 through 1977

Decision to Locate:	Number of Projects	Percentage
In districts of sitting members of the House Appropriations Subcommittee	7	4
In states of sitting members of the Senate Appropriations Subcommittee	55	34
Jointly in both districts and states of sitting members, Senate and House	16 (15)[a]	10 (9)[a]
Not located in districts or states of sitting members	82	51
Total	160	99

SOURCE: Agricultural Research Service (USDA)

[a] Numbers in parentheses indicate that 15 of the 16 joint projects were placed in House Chairman Jamie Whitten's district, and these 15 constituted 9% of all 160 projects.

ceived most of the construction that was jointly shared by House and Senate members.

That the House is a comparatively minor participant in pork barrel politics is evident from table 7.5, which shows that the districts of House members in 1977 and 1981 did not have many installations. Principal exceptions were Chairman Whitten's district, which gained a number of installations over three decades, though some were lost when district lines were moved following reapportionments, and Republican senior member Mark Andrews's at-large North Dakota district, which contained the land-grant university and several federal installations obtained mainly by the long-time senator from that state, Milton Young.

Senate pork-barreling has been a mixed blessing for agricultural research. The pork barrel impulse has been useful in gaining support for new activities, each of which then becomes an increment in total research funding. But the demand for state laboratories has

TABLE 7.5: Research Installations in Member Districts
of the House Appropriations Subcommittee on Agriculture, 1977 and 1981

Member and District	Research Installations and (Experiment Stations)
Democratic 1977	
Whitten (MS 1)[a]	Sedimentation Laboratory, Oxford
Natcher (KY 2)[a]	None
Evans (CO 3)	None
Burlison (MO 10)	None
Baucus (MT 1)	Rangeland Insect Lab, Bozeman
	Cereal, Forage Improvement
	(U. of Montana)[b] (Montana State U.)
Traxler (MI 8)[a]	None
Alexander (AR 1)[a]	None
Sikes (FL 1)	None
Democratic 1981	
McHugh (NY 27)	U.S. Plant Soil & Nutrition Res. Lab (Cornell U.)
Hightower (TX 13)	Guar Res., Vernon
	Gt. Plains Res., Bushland
	Amarillo Res. Ctr., Amarillo
Akaka (HA 2)	Sub. Tropical Fruit Insects Res., Hilo
Watkins (OK 3)	Ag. Water Quality Management Res., Durant
Republican 1977	
Andrews (ND, At-Large)	N. Gt. Plains Res. Lab., Mandan
	Human Nutrition Res. Lab., Grand Forks
	Met. & Rad. Res., Fargo
	Cereal Genetics, Path., Fargo (N. Dakota State U.)
Robinson (VA 7)[a]	None
Myers (IN 7)[a]	None
Republican 1981	
Smith (NB 3)	U.S. Meat Animal Res. Ctr., Clay Center
Lewis (CA 37)	Citrus & Date Prod. Research, Indio

[a] Indicates a member of the subcommittee in both 1977 and 1981.

[b] State Agricultural Experiment Stations.

122

obliged the federal Agricultural Research Service to operate a "traveling circus," opening new locations in current Senate constituencies, while closing some in states whose senators are no longer members of the subcommittee.

Another of the Senate subcommittee's low-information strategies has been to offer itself as a vehicle for outside expertise, especially the expertise of the administration and the research agencies themselves. Indeed, Richard Fenno found in his study of the congressional appropriating process that the Senate Appropriations Committee is a generous "court of appeals," through which the administration is able to salvage some of the funds deleted by the House.[3] For example, in 1977 the administration put forward a "new agenda" of research priorities. These priorities had been prepared in some cases by the research agencies themselves, but justification and support for this agenda had emerged also from the 1977 research statute, which had been supported by a broad coalition; from the earlier presidential commission of distinguished scientists which had produced the World Food and Nutrition Study; from an influential nutritionist, Jean Mayer; and from other authoritative sources.

An observation upon the ideology of recent Senate subcommittee members helps to explain why the committee added new agenda items to the research budget in 1977 (see table 7.3 for a list of these items). It can be surmised that the Senate subcommittee, in its support of "expert" members and home-state interests, was displaying a quite customary generosity toward agricultural clientele. But by supporting new agenda items, the group was also representing some new constituencies, including consumerists with liberal or even radical political associations. The Senate group supported this "liberal" new agenda again in 1978 and 1979. Thus the Senate subcommittee was at that time both generous and liberal in its orientation. The 1975–76 Senate membership, on the whole, was indeed ideologically liberal, as measured by the *Congressional Quarterly*'s scores on support for the Conservative Coalition of Republicans and Southern Democrats, and especially by comparison with past Senate membership and with the coterminous membership of the House subcommittee on agriculture. In 1975–76 the Senate Democrats' average score was 36 percent, as compared with the 1959 average of 64 percent. Similarly, the Republican average was 67 percent in 1975–76, 81 percent in 1959. In 1980, however, the Senate subcommittee had turned toward the conservatism of former years, with scores of 50 percent for the Democrats and 83 percent for the

TABLE 7.6: House Appropriations Subcommittee for Agriculture Membership 1959 through 1977

			Previous Members (1958–1980)		
Democrats	Year Began	Years Served	Republicans	Year Began	Years Served
Marshall (MN)	1953	9	Anderson (MN)	1945	12
Santangelo (NY)	1958	5	Horan (WA)	1945	14
Addabbo (NY)	1963	2	Vursell (IL)	1953	1
Hull (MO)	1965	10	Michel (IL)	1959	16
Morris (NM)	1965	4	Langen (MN)	1965	6
Evans (CO)	1969	10	Edwards (AL)	1969	2
Shipley (IL)	1969	8	Andrews (ND)	1971	9
Passman (LA)	1975	2			
Casey (TX)	1975	2			
Burlison (MO)	1975	6			
Baucus (MT)	1975	4			
Sikes (FL)	1977	1			
Jenrette (SC)	1979	2			

			1981 Membership		
Democrats	Year Began	Years Served	Republicans	Year Began	Years Served
Whitten (MS 1)	1945	32	Robinson (VA 7)	1971	11
Natcher (KY 2)	1953	24	Myers (IN 7)	1975	7
Traxler (MI 8)	1977	1	Smith (NB 3)	1981	1
Alexander (AR 1)	1977	1	Lewis (CA 37)	1981	1
McHugh (NY 27)	1979	3			
Hightower (TX 3)	1979	3			
Akaka (HA 2)	1981	1			
Watkins (OK 3)	1981	1			

Republicans. Average scores for the House subcommittee in 1975–76 were 50 percent for the Democrats, 88 percent for Republicans.

As indicated in the congressional staff survey (table 7.1), the House Appropriations Subcommittee on Agriculture wields major—some would say irresistible—influence over research decisions. It is a "one-man" committee, the man being its chairman for several decades, Congressman Jamie Whitten (D., Miss.) who is now also chairman of the full Appropriations Committee. Table 7.6 reveals that, as of 1977, Mr. Whitten had spent more years on the subcommittee than all other members combined, if one excludes his relatively silent longtime colleague, Congressman William Natcher (D., Ky.). Mr. Whitten's role as principal research decision maker is evidenced in the proceedings of his subcommittee, and is not disputed.

One task of the subcommittee, in the expectation of the present House, is "to make clientele-oriented decisions."[4] The subcommittee has another job assignment as well, however: to "guard the treasury through budget reductions."[5] In exercising this mandate the subcommittee serves as a high court among its clientele. It is expected to grant those research requests which meet a real need, which are technically sound and feasible, and—preferably—which have political support. Speaking through Mr. Whitten, the subcommittee has often rejected requests that fail on one or more of these criteria, as in 1977 it rejected big budgets for control of brucellosis and eradication of the boll weevil because the chairman could not be assured that these programs would accomplish their objectives.

In reconciling the "contrary" expectations of saving money on the one hand while satisfying clients on the other, the subcommittee becomes the decision center which chooses and oversees research programs serving a complex industry. Mr. Whitten's long tenure has provided opportunity for realistic time frames for planning and evaluating research. Over many decades the subcommittee's decisions have molded research institutions in the way the rule of *stare decisis* has molded law.

There are mixed opinions among industry insiders as to how well the subcommittee has performed. Some observers recall the "good" decisions. Others recall "bias" and "politics" in the subcommittee's decisions.

Table 7.7 is an effort to summarize the subcommittee's decisions or recommendations over a period of twenty-eight years, as these were highlighted in the subcommittee's reports on appropriations bills. We should keep in mind that the subcommittee was seeking in these House reports to give a sketch of its actions rather than an exhaustive description, and to do so in a context which would reassure potential critics in the House. Therefore, we must interpret these findings against the background of the hearings and by comparison with Senate and conference committee reports.

The subcommittee's role in responding to the immediate concerns of agriculture is evident in table 7.7. More than one-third of the subcommittee's "favorable" redirections were toward research on particular commodities or on bothersome pests and diseases. Among decisions concerning research on production or utilization of particular commodities, some "favorites" appear—cotton, tobacco, sugar, peanuts and oil crops, and fruits and nuts.

There were also negative redirections of some commodity research. In the hearings, the subcommittee showed its readiness to terminate the less productive research projects, and to disapprove some agency initiatives. From a comparison of House and Senate reports, it is also evident that the House subcommittee's negative redirections often were on items currently being championed by the Senate and likely to be restored during conference committee bargaining.

On first glance, table 7.7 would seem to indicate that the House subcommittee between 1950 and 1978 pushed for many of the "new agenda" objectives that have been more recently advocated by environmentalists and other nonindustry groups, and that it also fostered basic research of the kind advocated by panels of scientists between 1975 and 1977. The data assembled here may overstate the subcommittee's interest in new agenda objectives, however, because in past years research on these subjects served other objectives. The soil conservation research category, for example, comprises virtually all of the subcommittee's references to soil research, including research which was supportive of new technology packages designed to gain more efficient use of fertilizers and irrigation.

The subcommittee gave some support to the development of research infrastructures, particularly the germ plasm storage or seed banks which provided the genetic material for plant breeding. There was little mention of "basic research" per se, and a mixed response to demands for photosynthesis and nitrogen fixation research. The subcommittee did increase funding for research at the "1890s" colleges (the "predominantly black" land-grant colleges), partly in response to the concerns of one member, Congressman Frank Evans (D., Colo.).

Generally, the subcommittee discouraged new-agenda research, both because it was presumed to siphon resources from production research and because its findings sometimes challenged commercial practices. Human nutrition research was suspect because of the possibility that its findings might lead to reduced consumption of some food products; thus in years of tight budgets the subcommittee usually cut nutrition research more than it cut production research. Rural development research was opposed because it tended to promote interests other than those of commercial agriculture, including those of small towns, farm workers, and the rural poor, and because rural development research findings sometimes seemed to undermine industry interests, as in pointing out adverse social consequences of prevailing technology.

TABLE 7.7: House Action on Administration Budgets, 1950–1978

Research Categories	Budget Change[ab]		Approve Admin. Change[c]		Instruction Commentary[d]		Total Recommendations	
	+	−	+	−	+	−	+	−
New Agenda:								
Soil conservation	13				8		21	
Biolog. pest control	5		2		3		10	
Environmental preservation	4	1	1		3		8	1
Basic (photosynth., nitrogen fixation)	3	2					3	2
Germplasm (seed) banks	1		1	3	4		6	3
1890s colleges	5				3		8	
Direct marketing	1						1	
Rural development	3	8			1	5	4	13
Small farm		2			1	1	1	3
Human nutrition	4	2		1	5	3	9	6
Total	39	15	4	4	28	9	71	28
Economics:								
Cost-of-production			2				2	
Econ. investigation analysis		2						2
Marketing, retail, wholesale	1	2	1	1	2		4	3
Crop, livestock estimates	1	3		1			1	4
Total	2	7	3	2	2	0	7	9
Commodities:								
Cereals	3		5		1		9	
Sugar, sugar beets	5		1		1		7	
Oil, crops	7		1		2	1	10	1
Vegetables, potatoes, beans	3		5		1		9	
Fruits, nuts, bees	6		4		6		16	
Forage, pasture	1	2	1		1		3	2
Beef, pork, other	3	2	2	1		1	5	4
Dairy, poultry	2	2	1	1	2		5	3
Fibers								
Cotton	4				3		7	
All other			1		1		2	
Tobacco	3		1		2		6	
Plant breeding and nutrition	2			1			2	2
Ornamentals			2				2	
Total	39	7	24	3	20	2	83	12

TABLE 7.7 continued:

Research Categories	Budget Change[a][b]		Approve Admin. Change[c]		Instruction Commentary[d]		Total Recommendations	
	+	−	+	−	+	−	+	−
Utilization:								
Cotton	6		1		2		9	3
Other, general	10	1	2	1	7	2	19	4
Total	16	1	3	1	9	2	28	7
Pest Research:								
Specifically pesticides	5		2		3		10	
Pesticide regulation	4		2				6	
Other, general	20	9	5		6	1	31	10
Total	29	9	9	0	9	1	47	10
Plant and Animal Diseases:								
Animal diseases								
Brucellosis	8	1	5		4		17	1
All other animal diseases	18	4	2	1	7		27	5
Plant diseases	9		2	2	6		17	2
Undifferentiated animal-plant disease	2	5		1			2	6
Total	37	10	9	4	17	0	63	14
Other:								
Mechanization	2		3		2		7	
Hydrology, water, remote sensing	3						3	
Total	5	0	3	0	2	0	10	0
Grand Total	167	49	55	14	87	14	309	80

SOURCE: Reports on regular agriculture appropriations bills by the House Appropriations Committee, for budget years 1950 through 1978, including appropriations for Agricultural Research Service (previous to 1952 for the various USDA research bureaus), the Economic Research Service (from its creation in 1962), Cooperative States Research Service (and predecessor agencies), and the Extension Service.

a Plus symbol signifies a recommended addition. Minus symbol signifies a recommended reduction.

b Statements and listing of items indicating House changes in funding from those in the administration's budget, or House initiative in adding research projects.

c Statements indicating House approval of administration changes or addition of projects from the previous year's budget.

d Comments or instructions in the report relative to particular research programs or projects.

The subcommittee's judgment of industry interests was, of course, made in the light of its members' values. Those of its chairman, Mr. Whitten, are of particular relevance.

Chairman Whitten

Jamie Whitten was an early achiever.[6] He was born in 1910, in the small town of Cascilla, Mississippi, and at age twenty served one year as a school principal. He became a state legislator at age twenty-one, and a prosecuting attorney at age twenty-three. He became a congressman in 1941 at the age of thirty-one. He joined the appropriations subcommittee on agriculture in 1942, and in 1949 became its chairman. He has been chairman since that time, except during one Congress (1953–54) in which Republicans were in the majority.

Several indicators of his values are presented here. *Congressional Quarterly*'s analysis shows him voting with the House conservative coalition on 87 percent of its votes in 1959, 81 percent in 1975–76, and 76 percent in 1980. As another indicator of Mr. Whitten's interests, the record of bills introduced by him during his career has been analyzed. Parenthetically, aside from the annual appropriations bill, virtually none of these bills deals with agricultural research. Whitten's views can also be identified by reference to his statements during testimony before his committee.

District interests comprise one set of Congressman Whitten's concerns. Mr. Whitten has represented greatly varying portions of northern and western Mississippi in both the First and Second Congressional Districts of Mississippi, as boundaries have changed over the decades. His district has usually included both rich delta lands with large individual landholdings and a larger rural constituency with farms of low value. There are many farm towns, and also some suburbs of Memphis, Tennessee. About one-third of the population was black in 1960.

Whitten's bills expressing district interests include a number in behalf of Mississippi Delta landholders: he has sponsored legislation exempting them from flood insurance; permitting juries to set the value of lands purchased under eminent domain; reclaiming for private owners the mineral rights on federally purchased land; and providing for the resale of "surplus" land from that acquired by the federal government.

Some of Whitten's positions on research policy can also be attributed to his effort to serve district interests. Whitten participated in "pork barrel" research politics in order to ensure that a share of research funding would go to Mississippi's experiment stations. One researcher noted that in the funding of research against the 1970 epidemic of Southern corn leaf blight, "funds went to some states because legislators from these states sat on the agricultural committee. Mississippi was selected because of political pressure—Representative Whitten. We go along with Mississippi, or no money for anyone."[7]

On social welfare issues, Mr. Whitten's positions have been like those of many other rural conservatives representing low-income districts. He has been indifferent to or opposed to "welfare" measures such as food stamps. During his thirty years of service, the few welfare bills he has introduced or cosponsored have largely been intended to provide benefits to military veterans and their widows and children, or to extend coverage and benefits under the old age and survivors insurance (social security) program. He also introduced a voluntary health care insurance program for low-income persons.

Small farms, Mr. Whitten has argued, are economically obsolete, though he has approved them as places where some families could raise some of their own food.[8] He has opposed measures for protection of farm workers. In 1962 he expressed misgivings that a federal school for tractor drivers might serve as a means for unionizing them. Speaking indirectly of unions, he said, "I have personally always been afraid to let such control get into the production of food."[9]

On foreign affairs, Mr. Whitten has been an unpragmatic anticommunist—opposing agricultural trade (and loans) with communist and "nonmarket" countries. He has been a vigorous nationalist, demanding in a 1967 resolution that the United States maintain its sovereignty over the Panama Canal. In 1968 he introduced a resolution demanding that France begin paying her World War I debt to the United States.

Congressman Whitten has championed a number of "way of life" values. He has proposed several constitutional amendments to override landmark Supreme Court decisions: amendments to prevent "forced" busing to achieve integration; to allow prayer in schools; to recognize the authority and law of Jesus; to give the states exclusive control over health, morals, education, and marriage; and to permit the states to ignore the "one man, one vote" rule as a basis for distributing seats in one house of the legislature. He has also sought to create a standing committee in Congress whose function would be to "re-establish constitutional principles."

THE SUBCOMMITTEE & THE SUBSYSTEM COALITION

Over the years, Chairman Whitten's conservative values have given thrust to the subcommittee's role of organizing research for the agricultural industry while excluding other research agendas.

In writing research budgets the subcommittee has had the task of conciliating the interests of various research institutions, politicians, and farm groups in the search for consensus within the industry. For many years decisions were made by this political subsystem with little interference and even little attention from the president, the body of Congress, the media, and nonfarm groups. Whitten and other leading participants took pains to maintain this exclusive control. Infrequent challenges to the industry agenda came mainly from researchers within the subsystem who were promoting a technology that seemed to threaten a segment of the coalition, as oleomargarine technology once threatened dairy producers,[10] or who wished to probe the social and environmental side effects of the existing production technology, or who simply wanted to do research that was not of interest to any major segment of the industry. As compared with the state level, where research institutions have often been caught in a crossfire between commodity or industry competitors, a certain orderliness has been imposed in the national appropriations process. Mr. Whitten has maintained a network of cooperative administrators and informants. He has gained a reputation for knowing the details of research projects, and also for knowing what individual scientists are thinking. He has delineated areas in which he prefers that research should not be done, and has specified approaches or conceptualizations which he considers to be "wrong thinking." Notwithstanding the institutions' cleverness in resisting political direction, Mr. Whitten's wishes are likely to be heeded. In recent years, researchers and administrators have found themselves pressed to ignore Whitten's preferences because of mandates from the law or from the secretary of agriculture, or because Whitten's strictures were denying them new sources of research support, or simply because they have thought Whitten's ideas were wrong. In these cases they have acted at their peril. The safer tack—and certainly the more comfortable one for most research administrators—has been to satisfy the chairman. Although it is sometimes difficult to interpret the chairman's circumlocutory language—"he takes you all the way around the barn," said one administrator—there are instructive cases, vivid in the memories of administrators, as to what happens when the chairman's strong desires are misread or unheeded. Two such cases are presented here for illustration: the first, which occurred over the first two decades of Whitten's service, resulted in severe constraints upon social research; the second, which took place in 1972, alerted the new administrator of the Agricultural Research Service to the fact that the chairman expected him to prevent the expression of certain environmental viewpoints within the agency.

Banning Social Research

At the end of World War II, an ambitious USDA agency, the Bureau of Agricultural Economics (BAE), set out to deal in a comprehensive way with the problems of rural America, including the severe dilemmas of poverty-stricken, obsolescent sharecroppers and of the Negro war veterans returning to the rural South. Studies were made of 370 rural counties, from which the BAE was to aggregate information in making recommendations. One set of resulting recommendations was embodied in a "conversion plan for the cotton South," which recommended moves toward industrial development and agricultural diversification. The plan anticipated by about two decades the economic basis of the "new South." At the time, however, the BAE's approach was seen by Whitten and by most other rural southern congressmen as a threat to the societies they represented. It was branded by Whitten as "communist" and "socialist." He and others began a coalition to destroy the BAE as they had earlier destroyed the Farm Security Administration, an agency designed to help poor farmers and farm workers.

One of the BAE's 370 county studies was of Coahoma County in Mr. Whitten's district. A few paragraphs of the Coahoma County report, with unflattering references to racial segregation in the county, had found their way to Congressman Whitten and to the subcommittee's chairman at that time, Malcolm Tarver (D., Ga.). In cross-examining the BAE's chief, Howard Tolley, Mr. Whitten charged that the BAE had "tried to slander a fine people."

Mr. Tolley responded that the county studies had not been meant for publication but rather had been intended to provide internal information. But Mr. Whitten pressed him: "Do you not think that we would be doing the American farmer good if we were by legislation, if necessary, to put your bureau back to gathering agricultural statistics, and to take you out of the socialization field and the accumulation of claimed data and the printing of such vicious attacks on a county and its people as is done by your bureau in the case before us?"[11] Tolley finally acquiesced, stating, "Reports of this kind will not be made anymore."

The appropriations subcommittee, with support from other agricultural leaders, continued to attack the BAE's orientation toward rural development rather than toward agricultural production per se. Secretary of Agriculture Clinton Anderson asked for Tolley's resignation. Shortly thereafter, the BAE itself was reduced in functions, and then was abolished in 1953.

In 1961 a successor agency, the Economic Research Service, was established by Secretary of Agriculture Orville Freeman, but with an admonition from Whitten to Secretary Freeman:

> I would say for the record that in past years we had some problems with the old Bureau of Agricultural Economics. At one time some of their social

studies and other things were, to say the least, not very popular up at this level. It looked to us as if those things were getting over into the policy fields, that perhaps some undue influence was exercised on action programs by the theorists and economic groups.

It reached the point where it was extremely difficult for the Bureau of Agricultural Economics to get proper financing through the Congress. I don't mean that that will be the experience in the future, and I don't say it in any way to attempt to upset your plans at all. But it is always good for all of us to read history. May I just suggest that reading a little history might keep our new bureau in the proper field of activity, if the feelings of Congress are like they were some years ago.[12]

Mr. Whitten's appreciation of history did not extend to appreciating the USDA's centennial history, prepared for publication in 1962. Whitten endeavored to quash this publication largely because it contained a favorable reference to the former BAE head, Howard Tolley. Whitten finally agreed to release 6,000 copies of this work.[13]

The Ian McHarg Lecture

The concept of ecosystem, as an aspect of the environmentalist ethic of the 1970s, suggests that man is embedded in a circle of biological interactions which he must respect in order to save himself. Mr. Whitten considered as "extremist" the impulse to preserve the environment at the cost of not being permitted to reorder it. In 1972 he criticized the administrator of the Agricultural Research Service for permitting a visiting lecturer, Ian McHarg, to extol the concept of ecosystem.

The occasion was the 1971 USDA-sponsored Morrison lecture series, which had previously featured distinguished scientists as well as celebrities. Ian McHarg was a well-known landscape architect who had also been a celebrity guest on national television talk shows. In his Morrison lecture he suggested, tongue-in-cheek, that anyone who persisted in disrupting the earth's ecosystem should be sent off in a space capsule, within which the traveler's excrement would be rechanneled through a vat of algae, which would in turn provide him with food and water. When the traveler came to understand that "that's the way the world works," he would be welcomed home: "We would say to him,

'Come on home! Enter into the warming spiraling arms of the earth's gravity, enter this green celestial sphere that is our home, the home of our origins and that place where we will accomplish our destiny. Excercise now your creative will, because you have learned that deference born of understanding.' "[14]

Whitten felt that such a viewpoint was dangerous and insulting to our agricultural system and, therefore, that the ARS should never have permitted its expression. At the 1972 hearings he questioned ARS administrator T. W. Edminister:

WHITTEN: I think you might explain to me how in the world you spent some of your money in publishing a pamphlet here named "Man: Planetary Disease." We discussed this with the Secretary [Earl Butz]. He was a little embarrassed, if I am any judge. . . .
   I want the record to show that our committee had no knowledge that any money was available for this use. . . . Can you explain to me who in the world authorized the publication of this pamphlet? How did it happen?   •
EDMINISTER: Mr. Chairman, Mr. Ian L. McHarg was the 1971 Morrison Memorial lecturer. This is one of a series of two lectures we have sponsored since 1967. These lectures were established to broaden communication in science and industry.
WHITTEN: Do you consider that the product of a scientist?
EDMINISTER: It is the product of a scientist.
WHITTEN: What is your definition of a scientist?
EDMINISTER: A scientist in this case is a man who has been a leader in the field of landscape architecture, in community design, and one who had expressed a concern with regard to the environment. He came to us very highly recommended as an individual who had a concern for the place of the environment in the living conditions for the American public.
WHITTEN: You have read this pamphlet?
EDMINISTER: Yes, I have, sir.
WHITTEN: He puts man at the low end of the totem pole. Did he submit that paper to you and do your rules require it?
EDMINISTER: No.
WHITTEN: Your answer gives me no confidence in the judgment of your predecessors. . . . You changed the environment. You therefore come under the gun of this fellow. Yet you advertise him in ignorance by publishing and distributing such a lecture.
EDMINISTER: We never considered we were censors of free speech.

136

WHITTEN: If it were your money I might agree it was your privilege. However, you used funds appropriated by the Congress which come from the taxpayers, it is about time you have censors.

The message to Edminister was clear, as were the implicit consequences of his ignoring it.

THE INDUSTRY COALITION & THE GRAND COALITION

While some subsystem leaders—notably Mr. Whitten, many research administrators, and Secretary of Agriculture Earl Butz (1972–76)—have tried to minimize the influence of consumer and other nonindustry groups (to be discussed in chapter 8 as the externalities / alternatives coalition), other agricultural leaders have turned to these groups for support in passing farm legislation. In framing the 1977 farm bill, Secretary of Agriculture Robert Bergland and the chairmen of the congressional agriculture committees—Congressman Thomas Foley (D., Wash.) and Sen. Herman Talmadge (D., Ga.)—made trade-offs with these groups, one of which was the commitment to a new research agenda.

The congressional appropriations process still resists this new agenda, even though it was adopted into law. Congressman Whitten's rejection of much of the new agenda has made his own situation difficult on occasion. When congressional Democrats challenged the seniority system in 1974, opposition to Whitten from environmentalists and others almost cost him his subcommittee chairmanship. To save his position at that time he was obliged to yield his subcommittee's control over the budget of the Environmental Protection Agency (EPA). Whitten has remained in a posture mainly of conflict with new agenda interests—goading the USDA into a position of confrontation with the EPA; seeking to exempt agriculture from occupational health and safety regulations; seeking partially to reverse the EPA's ban on DDT; seeking to modify the Food and Drug Administration ban on saccharin and other food additives; and seeking to prevent the growth of competitive grants intended for new agenda research objectives. Obviously Whitten and other members of his subcommittee favor channeling money through administrators accountable to them rather than through the panels of scientists provided for under the competitive grant arrangement, arguing, as in 1972, that less money would be wasted if the subcommittee were allowed to decide on the specific projects before funds were granted.[15]

Congressman Whitten also fought a losing battle against a series of reorganizations of federal research, beginning in 1972, that created the national planning staff and a number of regional offices, apparently thus reducing the subcommittee's access to research decision making. In the 1976 hearings, Whitten protested to Assistant Secretary of Agriculture Robert Long that, as a result of the reorganization, his subcommittee could no longer get the "information" it needed, because it was obliged to seek it from the agency's administrator. Whitten objected that the administrator was "the man with an engineering background trying to present the case for the boll weevil man," and a brief colloquy ensued:

LONG: He pulls it all together.
WHITTEN: We would rather do it ourselves.
LONG: We can't change our organizations to the point we don't have an administrator.[16]

Congressman Whitten passed a major test of power in 1979, when, despite opposition from some environmental groups, his Democratic colleagues honored his seniority by elevating him to the chairmanship of the House Appropriations Committee. Whitten also retained his chairmanship of the subcommittee on agriculture.

THE APPROPRIATIONS CONFERENCE COMMITTEE

At the neck of the funnel of money that controlled agricultural research policy in 1977 was an unadorned room in the U.S. Capitol in which, around tables arranged in a large rectangle, the conference committee—composed of all members of the Senate and House agricultural appropriations subcommittees—met for five hours to iron out disagreements between the houses over the $13 billion agricultural budget. About half of this time was devoted to the research budget, although federal research comprised less than 6 percent of the total. Larger programs got less attention because the appropriations subcommittees had less discretion over their funding. Under the $5 billion food stamp program and even under farm subsidy programs, the law specified who was eligible and how much each should receive. But the research budget remained flexible. Members of the appropriations subcommittees could decide which labs were to be created and where, which insects were to be attacked, and which research agencies would be given responsibility for the various projects.

Several of the senators present at the conference wanted something for their home states. Sen. Birch Bayh (D., Ind.) made it clear that he would insist that funds be included for a federal soil erosion laboratory to be erected at Indiana's Purdue University. "I am supposed to chair an important hearing this afternoon," Bayh said, "but I will remain with this committee until this matter is resolved."[17] After Congressman Whitten protested that the Purdue facility might duplicate an existing lab in Whitten's own district, calls were made during the lunch hour to gain assurance from research heads that this would not happen. Bayh received his lab.

Sen. William Proxmire (D., Wis.) sought for his state a dairy forage lab which had been denied him during four previous years; this year he got it. The conference was reminded that the Senate's majority leader (also a member of the subcommittee) wanted a lab in his remote West Virginia hometown, and that wish, too, was granted.

The major Senate claimant was Sen. Henry Bellmon (R., Okla.) whose ranching experience, expertise, and bull-like approach made him a formidable challenger to the House chairman. Bellmon wanted more funds for fighting the lonestar tick and the range land witchweed; new research on the tanning of leather; a soil laboratory for Oklahoma; and additional millions of dollars for a faltering campaign against brucellosis (a venereal disease in cattle). Whitten was opposed to increasing brucellosis funding, and was as stubborn on this issue as Bellmon was adamant. Said Whitten: "I started in 1949 as chairman and we've tried time after time to stop it and it hasn't worked." Bellmon countered that it would work with greater funding. Brucellosis funds and some other unresolved issues were set aside for bargaining at the end of the meeting. At the conclusion, the Senate's novice chairman, Thomas Eagleton (D., Mo.), suggested a package of compromises on unresolved matters, most of which were accepted by Mr. Whitten.

Mr. Whitten, usually a dominating presence at these conferences, was more accommodating than usual at this one. "Congressman Whitten is so sweet today, it is unbelievable," said one Senate staff member to another. An outsider lacking previous basis for comparison would have sensed toughness more than sweetness in Whitten's remarks on several subjects:

—*on the veracity of research administrators.* When told that Frank Mulhern, head of the USDA's antipest agency, had said that a brucellosis campaign could work, Whitten said: "Send for Mulhern. I'll get him to admit everything I told you. Mulhern is a very good man, but let me question him and I'll have him telling me everything I've said is true." In responding to someone who had obtained information from a federal research administrator, Whitten chided: "You'd better get a better witness. I've been listening to them a long time. They moved the ARS out of Washington so they wouldn't have to report to Congress. You've got to go to the field for the information."

—*on the uselessness of social science studies.* "They did a study of my town and I could have called down and got just what they found out after they had a bunch of people go down there."

—*on the boneheadedness of agricultural scientists.* "Sometimes we have to tell them what to do, because these scientists are inclined to do what they've been doing. They were studying swamp fever in mules after the mules were gone."

Mr. Whitten got the best of the bargain on the big research issue: the administration's request for $28 million to be used for competitive grants. These funds for new agenda objectives were to be distributed by panels of scientists rather than by Mr. Whitten's committee and agency administrators. The Senate subcommittee had approved the administration's request, but Whitten, opposed to this change in substance and procedure, had allowed only $10 million. Whitten suggested that if the senators would reduce the competitive grants, House members would let them "make it up elsewhere"; that is, the senators would be free to add their home-state "pork barrel" projects. The resulting compromise provided only $15 million in competitive grants. In the following year—1978—Mr. Whitten was again successful in holding down the competitive grants for new agenda items, which by then had been authorized in law; in 1979, with support from an industry coalition, his committee—and the conference committee—drastically reduced the administration's competitive grants budget.

COMPARATIVE IMPACTS OF SENATE,
HOUSE, AND ADMINISTRATION

Table 7.8 offers one test of various presumptions about relative support for research budgets within Congress and the executive. In this table, budget recommendations of the administration and the House and Senate subcommittees are compared over two decades to determine which of these decision makers tended to recommend the highest and lowest figures and also to indicate which of these recommendations "won out." The table provides data for four different agricultural research agencies: the Agricultural Research Service, the Economic Research Service, the Cooperative States Research Service, and the Extension Service.

The table affirms that the Senate subcommittee has embodied a generous spirit, very likely expressing an impulse of the Senate as a whole to support programs rather than cut budgets.[18] For three out of four research agencies (the exception is the Economic Research Service), the Senate usually recommended a budget larger than that which the administration had submitted.

Several results are not as may have been expected. The Senate subcommittee was particularly generous between 1957 and 1970, when its membership was quite conservative, as indicated earlier by the 1959 Conservative Coalition scores. Table 7.8 also reveals another apparent incongruity: during the earlier period, the final appropriation was

TABLE 7.8: Impact of the Congressional Appropriations Committees on Research Funding

Year	Highest + Lowest − Admin.	House	Sen.[a]	Final approp. was closest to the House	Senate[b]	Final approp. above and below admin. budget[c]
Agricultural Research Service						
1957	+		−		x	−
58	+	−		x		−
59	+	−			x	−
60		−	+		x	+
61		−	+		x	+
62			+		x	+
63		−	+	x		+
64	−	+			x	+
65		−	+		x	+
66	+	−				+
67	−		+		x	+
68		−	+	x		−
69	+	−		x		−
70		−	+	x		+
71			+	x		+
72	−		+	x		+
73	−		+	x		+
74	+		−	x		−
75		−	+	x		−
76	−		+		x	+
77	−		+	x		+
78		−	+	x		+
Economic Research Service						
1963	+	−		x		−
64	+	−			x	−
65		−	+		x	+
66		−	+		x	+
67	+	−			x	−
68	+	−		x		−
69	+	−		x		−
70	+	−	+	x		−
71	+	−	+	x		−
72	−	+	−		x	same
73		−	+	equal		−
74	−	−	+		x	+
75		−	+		x	+
76	+	−		x		−
77		−	+	x		−
78		−	+		x	+

TABLE 7.8 continued

Year	Highest + Lowest − Admin.	House	Sen.[a]	Final approp. was closest to the House	Senate[b]	Final approp. above and below admin. budget[c]
Cooperative States Research Service						
1957	−	−	+	x		same
58	+	−			x	+
59	−	−	+		x	+
60						same
61	+	−	+		x	−
62	−	−	+	mean		+
63						same
64	−	−	+		x	+
65		−	+		x	+
66		−	+		x	+
67	−		+		x	+
68	−	−	+	x		same
69	+	−		x		−
70	+	−			x	−
71	+	−			x	−
72	−		+		x	+
73	−		+	x		+
74	−		+		x	+
75	−		+	x		+
76		−	+	mean		−
77	−		+	x		+
78	−		+	x		+
Extension Service						
1957	+	−		x		−
58	+	−			x	−
59	−	−	+		x	+
60	+	−	−	mean		−
61		−	+		x	+
62		−	+		x	+
63		−	+		x	+

TABLE 7.8 continued:

[a] Plus symbol indicates the highest recommendation (or the identical ones tied for highest). Minus symbol indicates the lowest recommendation (or the identical ones tied for lowest).

[b] Letter "x" indicates which House recommendation was closest to the amount finally appropriated; "mean" indicates the appropriation was the mean of the House and Senate recommendations.

[c] Plus symbol indicates that the appropriation was more than what the administration requested. Minus symbol indicates that the appropriation was less than what the administration requested.

TABLE 7.8 continued

Year	Highest + Lowest − Admin.	House	Sen.ᵃ	Final approp. was closest to the House	Senateᵇ	Final approp. above and below admin. budgetᶜ
64		−	+		x	+
65	−		+		x	+
66	−	−	+	x		+
67		−	+		x	+
68	−	−	+	x		+
69	+	−			x	−
70		−	+		x	+
71	+	−		x		−
72	−		+	x		+
73	−		+	x		+
74	−		+	mean		+
75	−		+	x		+
76	−		+	x		+
77	−		+	x		+
78	−		+		x	+

ᵃ Plus symbol indicates the highest recommendation (or the identical ones tied for highest). Minus symbol indicates the lowest recommendation (or the identical ones tied for lowest).

ᵇ Letter "x" indicates which House recommendation was closest to the amount finally appropriated; "mean" indicates the appropriation was the mean of the House and Senate recommendations.

ᶜ Plus symbol indicates that the appropriation was more than what the administration requested. Minus symbol indicates that the appropriation was less than what the administration requested.

usually closest to the Senate's recommendation. This suggestion that the Senate was once more influential than the House conflicts with the bulk of evidence that the House subcommittee has been Congress's major decision maker on agricultural research.

The explanation may be that appropriations decisions are, for the House subcommittee, as much a source of power as an exercise of it. Decisions on appropriations may be anticipatory moves, as Aaron

Wildavsky has pointed out.[19] The House, usually starting with a low figure, can state the terms on which it will allow agencies an increment over the previous year's budget, an increment which the agency will already have constructed to include the favorite projects of various Senate members. For the knowledgeable Mr. Whitten, authority over budgets surely undergirds his more precise controls upon agricultural research. The senators, in contrast, are evidently unwilling to spend much time on research oversight.

The administration's impact upon research budgets appears to have varied, among agencies and over time. A comparison of the recommendations of administration, House, and Senate for the Agricultural Research Service shows that the administration budget was highest or in the middle of the three during the first decade, lowest or in the middle during the last decade. The appropriation for ARS has usually exceeded the administration's request, and therefore has not been in conformity with the norm that Congress is expected to cut administration budgets.

For the Economic Research Service, in contrast, the House figure was almost invariably lower than the administration's, and the impact of Congress was usually to reduce the administration's budget. For the Cooperative States Research Service (state experiment stations), House proposals were lowest until 1972, after which the administration's were lowest. From 1972, the Senate's experiment station budget was almost invariably the highest. For the Extension Service, the administration's recommendation has usually been lowest and has been raised by action of Congress. It can be concluded that the executive branch has recently been less supportive than Congress of assistance to state experiment stations and extension, and this has been true for both Democratic and Republican administrations. Congress has also been more generous with the Agricultural Research Service, while subtracting from the budget of the Economic Research Service.

Table 7.9 is an ambitious effort to state the predominant direction and intensity of change from previous years' budgets. The table was

145

constructed by calculating percentage changes for each agency or bureau from the previous year's budget as well as percentages by which Congress changed administration requests, then scrutinizing each set of changes to obtain an estimate of predominant direction and intensity of change.[20] Time periods coincide with changes of political parties within the administration, with the Roosevelt period being divided into war and peace periods.

In table 7.9, congressional action is measured by degree of concurrence with administration recommendations. For example, during the period 1969–76, when the Republican administration usually recommended increased budgets, congressional appropriations concurred with administration recommendations for three years, but in one year Congress appropriated much more than the administration had recommended. Two other administration budgets during this period were revised upward by 3 to 8 percent, while two were revised downward within the same range. Over the years most administration budgets recommended incremental increases. Republican administrations recommended large increases during five of sixteen years, while Democratic administrations proposed large increases in only three of twenty-eight years. Congressional action was likely either to be in concurrence with the administration's recommendations (whichever direction they took) or to revise those recommendations slightly upward.

COMMENTARY

Public research agencies have received much support from Congress, and indeed have looked to the congressional appropriations committees as principal supporters. Public research is an "old" governmental activity which grew up during the period of legislative control over budgeting and remains more or less in that mode, despite recent efforts to assert control from various levels within the executive branch. Legislative support has come largely from rural conservatives, who, until quite recently, have dominated the Congress. Rural legislators have wished to impose upon agricultural scientists an authoritarian relation-

146

TABLE 7.9: Action on Research Budgets: Congress v. Executive Branch

Action / Category	Periods					
	1933 to 1940	1941 to 1945	1946 to 1952	1953 to 1960	1961 to 1968	1969 to 1976
Executive Recommendation:[a]						
Large decrease (−9% or more)	2					1
Incremental decrease (−3 to −8%)	1	1	1		1	
No change (−2 to +2%)		1	3	1	1	
Incremental increase (+3 to +8%)	4	3	1	5	6	4
Large increase (+9% or more)	1		2	2		3
Congressional Revision:						
Large revision downward (−9% or more)						
Small revision downward (−3 to −8%)			2			2
Congress concurs (−2 to +2%)	7	4	3	6	5	3
Small revision upward (+3 to +8%)	1	1	2	1	2	2
Large revision upward (+9% or more)				1	1	1

SOURCE: Budgets of the United States.

[a] The general pattern of *change*, a judgment based upon observation of recommended changes for research bureau budgets (percentage of previous year's budget).

[b] General pattern of *revision* in executive budget, a judgment based on the differences between percentage change from previous year's budget as recommended by executive and percentage change from previous year budget as actually expressed in the next year's budget.

ship similar to that which farmers customarily imposed upon their children, their hired hands, and their wives.

Legislators, especially in the Senate, have often been motivated by the pork barrel—the desire to place buildings, jobs, and research missions within their own states. Pork-barreling is scorned as wasteful by modern budget managers, and this partly explains the low opinion of agricultural research held by some executive budget makers, even in the face of evidence that agricultural research has been extraordinarily cost-efficient. We should instead give considerable credit to pork barrel politics for several fortunate developments in agricultural research. In proliferating research locations, for example, Congress has helped unlock high-potential regions such as the High Plains. Moreover, pork barrel politics have sustained experiment stations in poor states, which in some ways has helped alleviate poverty and malnutrition there, as in the introduction of "greens" into diets.

Pork barrel has also, quite typically, permitted scientists to pursue long-range goals of their own choice which might not otherwise have achieved political support, an example being the human nutrition research laboratory located in North Dakota. The influence of pork barrel may also help to account for the extraordinary specialization at the state level, the benefits of which may largely accrue to the nation and world, an example being the research on wheat breeding, milling, and baking done at Kansas State University.

Because of the seniority system, and especially because of Chairman Whitten's long tenure, Congress has been able to provide the expansive time frames which the research institutions need. On the other side, the biases of senior legislators may impede good decision making. Says a veteran spokesman for the experiment stations: "I know everybody has a bias, but Jamie Whitten usually expressed his first, and that's what we had to do. It was usually not possible for us to talk things out until we had gained the best possible thinking as to what we should do." Whitten's biases have usually reinforced the prevailing notion, in research decision making, that research which serves the industry's interests also serves the nation well. The result has been a very narrow perspective. With all the ferocity of the classic "mad scientist," the creators of our agricultural revolution have put out of mind the massive social and environmental consequences of technological change. While most of the nation's black population was being displaced from southern farms into northern cities, efforts to understand this migration and ameliorate its effects were stifled. Researchers were denied freedom to observe the impact of agricultural technology upon the ecosystem. These heavy-handed restraints on the scope of agricultural research invited criticism from outside and, when the criticism came, produced a "fortress mentality." Thought and expression became defensive and conformist. For example, at a meeting of experiment station and extension service administrators, the group applauded exuberantly when one of their number announced (inaccurately, as it turned out) that Barry Commoner had lost his university laboratory. Commoner, who has explored the relationship between agriculture and the ecosystem, had been perceived as an enemy of the agricultural research establishment.

Those outside the industry who press research institutions to take on a new agenda—and the political leaders who press them to join with nonindustry groups into a grand coalition—are asking them to forsake a way of thinking and acting that still has considerable support within the congressional appropriations process.

CHAPTER EIGHT

The

Externalities / Alternatives

Coalition

A large "new agenda" for agricultural research has been imposed upon the research institutions by groups and agencies referred to here as the externalities / alternatives (ex/al) coalition. This chapter describes the activities and the impact of these ex/al groups, focusing on pest control research policy as a case study.

Figure 8.1 presents a model explaining the influence of ex/al groups upon the research agenda, using pest control research as the example. These groups supported research on pesticide hazards, and they also advocated research on alternatives to chemical pesticides. Their fundamental impact, however, was in expanding public concern over hazards from pesticides: hazards to those who consumed residues in food; hazards to applicators and others who came into contact with pesticides; hazards also to the natural environment. The objective of ex/al groups was to generate public support for pesticide regulation, because the prospect or existence of regulatory constraint—for example, that under which DDT was banned—helps to move industry groups to support research on alternative pest controls. In this roundabout way, also, ex/al groups are able to affect the research agenda.

Another unintended result of regulatory law, and also of research on

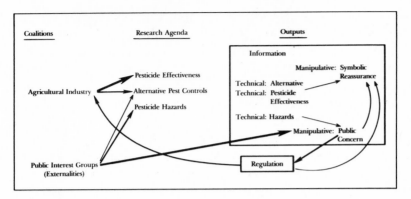

FIGURE 8.1: Information and Research Policy Processes.

alternative pest controls, is that an interested public may be reassured that the problem is being handled. Major questions to be considered in this chapter, therefore, are two. How do ex/al groups generate a concerned public? And, is this concern only fleeting, or does it persist?

PUBLIC AWAKENING

The pest control battle may fairly be said to have begun in 1962, with the publication of Rachel Carson's insistent book *Silent Spring*.[1] Carson imagined a farm on which human and nonhuman life had been blighted by deadly chemicals, resulting in a "silent" springtime, and her analysis of specific pesticide hazards crossed the threshold to a concerned public. Building on her success, ex/al groups channeled public concern into support of new regulatory law and a new Environmental Protection Agency.

Within the agricultural subsystem leaders had discouraged agricultural researchers from studying hazards and other externalities, as noted in the previous chapter, and they had endeavored to reassure the public that pesticides were safe as well as necessary. The endeavor continued even after *Silent Spring*, as Congressman Jamie Whitten wrote a book in answer to Carson's.[2] Now, facing the prospect that existing technology would be sidelined by health, safety, and environ-

151

mental regulations, leaders of the subsystem pressed ahead for safe pesticides, and also for alternative means of pest control.

When subsequent research did reveal pesticide hazards, several major pesticides were banned. Some alternatives were found, both in the form of acceptable new pesticides and in other methods of control which could be integrated with pesticide use. Although these responses by the industry and by the ex/al coalition apparently reassured the concerned public, it remained attentive to the issues of pesticide use.

The processes which created the new agenda and new policy evolved over a period of two decades. Both political change and research development require a good deal of time, and this fact adds complexity to the relationship between them. Several years were required to expand public and institutional concern about pesticide hazards to the point at which they would be given serious attention. Additional years were needed to enact and implement regulatory and research programs. More time was required for research which would examine claims that pesticides are dangerous to humans and the environment; and more still was needed for the development of safer yet effective pesticides and for the testing of pest control strategies which would place less reliance upon pesticides. The ultimate outcome represented a major success for environmentalist and consumer groups.

PRE-CARSON POLICY MAKING

Since the mid–nineteenth century, farmers have used pesticides on fruits and other high-value commodities, even though the early pesticides, such as lead arsenate, were toxic to humans and not very effective against pests. The developing science of entomology had the task of finding chemical concentrations more toxic to insects than to man. During World War II the insecticide DDT was developed in European industrial laboratories. DDT and other chlorinated hydrocarbons were inexpensive, and effective against a wide spectrum of insects at concentrations apparently not toxic to man. But these chemicals persisted in the environment, and they accumulated (biomagnified) in the food chain, with the result that some birds and other animals died of DDT poisoning. Most organisms, including the human body, became reservoirs of DDT.

Agricultural researchers viewed pesticide technology as a boon to farming and helped farmers maximize its effectiveness. They also developed some workable alternatives to pesticides, such as crop rotation as a means of controlling corn rootworms and production of sterile males to control screwworms, boll weevils, and other major pests.

Pesticide policy was made by the industry subsystem, which framed a regulatory bill in 1947 from conferences among chemical companies, trade groups, farm organizations, and the Department of Agriculture, all of whom testified favorably on the bill at a round-table appearance before the House Agriculture Committee. Differences had to be resolved—for example, over whether there should be prior registration of commercial pesticides. Under the law as passed, registration was required, but products which were refused registration could be sold under "protest registration." This permissiveness was accepted by the USDA in order to assure industry support of the law.[3]

Clearly, the law was written to serve the values of the agricultural industry.[4] The industry wished to use pesticides for profit, for convenience, and for the status which comes to "good farmers" whose fields are clean and bountiful. The intent of the law was to facilitate pesticide use by penalizing those who misrepresented or misused pesticides.

The industry coalition, by controlling production of expert knowledge and access to it, as reviewed in the previous chapter, tried to maintain control over pesticide decisions. Subsystem leaders tried to assure others that public interests were being served. They used the positive imagery of cheap and abundant food, and also the imagery of an agricultural structure based upon relatively small but innovative family farms. Pesticides were presented as effective foes of such "vile creatures" as malaria-bearing mosquitos, ants which inflict painful bites, and rapacious worms.[5] Pest research was designed to enhance pest control effectiveness. Agricultural politics characteristically sought to justify the industry's exclusive jurisdiction, while generating consensus within the industry.

153

THE EX/AL GROUPS

Groups challenging the agricultural establishment used a contrasting imagery. Rachel Carson's *Silent Spring* envisioned the death of wildlife, and even of the environment that sustains human life. Carson blamed a mistaken expertise: entomologists who sought "control of nature," she said, were from the "Neanderthal Age of biology and philosophy." She concluded: "It is our alarming misfortune that so primitive a science has armed itself with the most modern and terrible weapons and that in turning them against the insect it has also turned them against the earth."[6] Subsequently, ex/al groups accused agricultural policy makers of being subservient to the interests of large corporations, and of participating in a "mafia" which exercised a heavy hand against those who dared discuss the hazards of pesticides.[7]

Those presenting this new imagery were mainly groups whose intent, like Rachel Carson's, was to protect the natural environment. Environmentalist groups are often classified as "public interest groups," that is, groups which seek a "collective good" rather than the furtherance of their own interests.[8] In 1977 Jeffrey Berry identified eighty-three such organizations operating in Washington, of which one-fourth were environmental groups. Table 8.1 contains the public interest groups which were identified by Berry, and adds others which were not on Berry's list but which testified at one of three congressional hearings. A considerable number of environmental, consumer, and church groups did testify at the House Agriculture Committee's hearings on the 1977 farm bill, but only two of them thought it worthwhile to testify at the House agricultural appropriations hearings, where major research decisions are made. Among those public interest groups which appear in the *Encyclopedia of Associations,* only a minority indicate an interest in research. The major thrust of environmental and consumer groups has been to demand regulation rather than research or other subsidy.

Berry learned, from interviews with public interest groups, that most of them have been created at the initiative of an individual leader or "entrepreneur," rather than in response to some crisis or disturbance.

About one-third of them have no membership, and another third have fewer than 100,000 members. The few with large memberships, such as the National Audubon Society, have attracted members by offering nonpolitical benefits (for example, a subscription to an illustrated periodical). Typical methods of securing funds are mail solicitation of acquired lists of potential donors and support from private foundations. Mail solicitation succeeds when pegged to public apprehension, and responses decline when the public becomes reassured that the problem is being resolved. To stay in existence, groups are obliged to seek new issues as steppingstones to public support, or else to rely upon grants from foundations or from sympathetic governmental agencies.

Most public interest organizations have small budgets and only two or three full-time professional staff members, with occasional assistance from volunteers. In contrast with conventional Washington lobbies, as described in Lester Milbrath's earlier study,[9] decisions on group stands are made by staff members rather than by the membership or by elected boards. Staffers are for the most part young, and personally committed to the cause they represent. Berry's public interest representatives seek to influence both government and attentive publics.[10] They use some conventional lobbying techniques, including letter-writing campaigns, contact with sympathetic legislators, and presenting testimony at legislative hearings, but they are quick to use political protest and other unconventional techniques to gain public and official awareness of an issue and to make themselves visible. Another way of attracting attention—particularly from the media—to an issue is to produce a sensational report, *Silent Spring* being an outstanding example. By providing eye-catching information—"news stories"—on matters about which readers can develop high concern, the public interest group seeks to benefit from media influence upon popular perceptions.

After drawing attention to an issue, these groups generate a continuing controversy around it. For example, they embarrass and confront officials in the hope that the latter will become combatants.[11] They also pursue an inside strategy of providing information to friendly

The Externalities / Alternatives Coalition

TABLE 8.1: Public Interest Groups in Agriculture Policy

| | Testified[a] | |
	House Ag. Appro. 1976 1979	1977 Farm Bill
Environmental Groups:		
American Horse Protection Assoc.		
Citizens Committee on Natural Resources		
Defenders of Wildlife		
Environmental Action		
Colorado River Basin Salinity Control Forum		
Environmental Defense Fund	x	
(The) Environmental Fund		x
Environmental Policy Center		x
Fishermen's Clean Water Action Project		
Friends of the Earth		
Funds for Animals		x
Humane Society of the United States		
Izaak Walton League		
National Aububon Society		
National Resources Defense Council		x
National Wildlife Federation		x
Population Crises Committee	x	x
Rachel Carson Trust Fund for the Living Environment		x
Sierra Club		x
Society for Animal Protection Legislation		
Trout Unlimited		
Wilderness Society		
(The) Wildlife Society		
Zero Population Growth		x

TABLE 8.1 continued:

	Testified[a]	
	House Ag. Appro. 1976 1979	1977 Farm Bill
Consumer Groups:		
Ad Hoc Comm. for Rice Consumers (Wash.)		
Agribusiness Accountability Project		x
Aviation Consumer Action Project		
Center for Auto Safety		
Center for Science in the Public Interest		
Citizens Communication Center		
Committee for National Health Insurance		
Community Nutrition Institute		
Consumer Coalition		x
Consumer Federation of Earth		x
		x
Consumers Union		
Corporate Accountability Research Group		
Media Access Project		
National Consumers League		
Project on Corporate Responsibility		
Public Citizens Congress Watch		
Public Citizen Litigation Group		x

Church Groups		
American Baptist Churches, USA		
Bread for the World		x
Church of the Brethren		
Church World Service		x
Clergy and Laity Concerned		

157

TABLE 8.1 continued:

	Testified[a]		
	House Ag. Appro.		1977 Farm Bill
Church Groups	1976	1979	
Friends Committee on National Legislation			
Lutheran World Relief			x
Mennonite Central Committee			
National Council of Churches			
National Rural Housing Coalition	x	x	
Overseas Development Council			x
Program Dept. of CARE			x
Holy Family Church (Omaha, Nebraska)			x
Interreligious Task Force on U.S. Food Policy			x
Rural America (Inc.)			
Southern Rural Policy Congress		x	
United Church of Christ			
United Methodist Church			
United Presbyterian Church			

[a] The organizations listed here include a sample of public interest groups identified by Jeffrey M. Berry, listed in *Lobbying for the People* (Princeton, N.J.: Princeton University Press, 1977), plus others which appeared in one of the above hearings. Berry lists additional categories of public interest groups with the following numbers of groups in each category: civil rights and poverty, 5; peace, 15; general politics, 17; and miscellaneous, 2. No groups in these additional categories were active at the agricultural hearing.

legislators and administrators, but they stand ready to censure "friends" as well as "enemies" on the assumption that neither likes unfavorable publicity. These groups bring lawsuits which require officials to act or which give them an excuse to do so.

By generating information, imagery, and demands for action, the public interest groups provide news for the media, mandates for agencies and committees, and issues on which politicians can assume leadership roles. For example, public concern led to the establishment of the Environmental Protection Agency, which in turn has produced regulations constraining agricultural technology and which also has organized and subsidized research on pesticide hazards. Existing regulatory agencies such as the Food and Drug Administration, the Public Health Service, and the Occupational Safety and Health Agency have found new mandates, from law or, indirectly, from public concerns, relative to the agricultural industry. Federal developmental agencies such as the Agency for International Development have made policies affecting the direction of agricultural research. Within the federal government, other major research establishments, including the National Science Foundation and the National Institutes of Health, have become involved in doing agricultural research because of such "externalities" concerns as the impact of food upon nutrition and health and because some members of the larger scientific community have desired to explore alternative approaches to agricultural production.

POLICY SEQUENCE EVIDENCED IN THE MEDIA

The progression from public concern to public policy is evidenced in media coverage of pesticides since 1963, the time at which the subject had begun to claim major attention. Table 8.2 tallies the number of articles on pesticides appearing in the *New York Times* and the *Wall Street Journal* for the period 1963 to 1977. There are three periods of extraordinarily heavy media coverage of pesticides: 1963–64; 1969–72; and 1976–77. During the first period, following the publication of *Silent Spring*, there were demands for action, increased coverage of reports from monitoring agencies, and increased reports of pesticide damage. President Kennedy's Science Advisory Committee did a special study which recommended improved monitoring of dangers and more research into the hazards of chemicals as well as alternatives to them, along with a strengthening of the regulatory law.

During the following four years, however, there was a remarkable

The Externalities / Alternatives Coalition

TABLE 8.2: Public Exposure to Pesticide Issues: Articles in the *New York Times* and *Wall Street Journal*

Year	1963	64	65	66	67	68	69	70	71	72	73	74	75	76	77
Total Articles[a]	70	71	37	49	32	41	149	147	102	97	64	93	89	153	116
Reports of Damage, Hazards	16	17	7	9	7	5	17	15	19	7	3	14	7	21	7
Government Regulatory Activity[b]	7	9	2	3	1	5	24	19	16	26	11	14	19	15	7
Support for Pesticides[c]	7	6	8	2	2	0	8	11	5	6	6	2	4	4	3
Demand for Regulation	3	21	3	3	1	5	37	18	18	9	2	3	6	9	6
Gov't. *Use* of Pesticides[d]	1	0	1	1	0	0	6	2	4	3	3	1	6	1	1
Ongoing Information Monitoring	11	12	5	2	1	3	3	8	2	4	2	3	7	10	9
Activity in Courts	1	0	2	2	6	3	3	9	4	5	0	6	3	14	5
Private Firms Positive to Pesticides	2	0	1	0	0	3	1	0	1	0	0	0	2	5	2
Private Firms Negative to Pesticides	0	0	0	0	0	1	6	5	1	1	1	0	0	2	0
New Technology	9	5	2	11	8	7	7	17	8	6	2	4	7	23	6

NOTE: Articles are classified here as referenced in the indexes of the *New York Times* and the *Wall Street Jounal.*

[a] Because some articles were not classifiable, the number of "total articles" is larger than the sum of the categories.

[b] Government law making, rule making, implementation.

[c] Repudiating claims of damage, citing pesticide benefits, opposing regulation.

[d] Government administers pesticides or recommends their use.

reduction in media attention to pesticides issues. Time may have been required for new research to be produced, for public awareness to grow, and for the unfolding of political processes which would contribute to an outburst of media coverage in 1969. During this time, extended hearings were conducted by a subcommittee of the Senate Government Operations Committee, under Senator Abraham Ribicoff (D., Conn.), which in 1966 made a report calling for a stronger law and for more interdepartmental cooperation.[12] In 1965, Congress authorized large increases in research funding for the USDA, and also for HEW and Interior, for studying the "effects of pesticides." The Johnson administration submitted to Congress a pest control regulatory bill, but neither of the congressional agriculture committees acted upon it. In 1968, a General Accounting Office study pointed up serious deficiencies in the USDA's regulatory administration, and this report was followed up, in 1969, with hard-hitting hearings by the House Government Operations Committee.[13]

Between 1965 and 1969 there were many research reports assessing pesticide dangers, including a much-cited summary of carcinogenicity findings.[14] In addition, pesticide effectiveness research was proceeding, and pesticide use was increasing. Environmental contamination from DDT and kindred chemicals was becoming more evident and more widespread. Meanwhile, the USDA continued air spraying of DDT in campaigns against insects, seemingly unaware that many people in the areas being sprayed had become involved in pesticides issues, and might react strongly to a personal confrontation with pesticides.

As a result of all these activities, pesticides became an inviting "buzz word" on the agenda of many legislators and agencies. Actions taken by the government led, in turn, to further expansion of information and concern. Because of the broad-based public concern over pesticide regulation, as evidenced in table 8.2, pesticide policy would no longer be decided within the agricultural committees; instead, regulation was transferred to a new agency, the Environmental Protection Agency. Over the next several years a number of congressional committees, including the Senate Commerce Committee, and several executive agencies, including the EPA, took actions designed to alleviate public concern about pesticide hazards. These actions included passage of a new law and the banning of several major chemicals.

RESEARCH ON PESTICIDE HAZARDS

Table 8.3, tallying U.S. publications on pesticides since 1948, is offered as an index to federal support of pesticide research. During the 1950s, a large federal output on pesticide effectiveness overshadowed a relatively small output on pesticide hazards. After 1963, however, research on hazards was more proportionate to that on effectiveness.

The late arrival of hazards research is indicated also in table 8.4, which cumulates by year of publication the hundreds of citations of previous research which appeared in the 1969 report of the Mrak Commission, appointed by the secretary of Health, Education and Welfare to study pesticides and their relationship to environmental health. The small number of citations from years before 1960 suggests that Rachel Carson's charges may have lacked scientific documentation, a point her critics stressed. Indeed, the staff of Congressman Jamie Whitten's House appropriations subcommittee, after interviewing more than two hundred "experts," reported them as having repudiated most of her charges on grounds either that "there is no evidence" or that they were "possibilities as yet unproved to be actual facts."[15] This "expert" repudiation based on the absence of research proved to be premature on such key questions as whether or not exposure to pesticides might cause sterility or cancer.[16]

Scientists in a few fields did register early and active concerns relative to the side effects of agricultural chemicals. Their activity is indicated by a few early publications cited in the report of the HEW Secretary's Commission on Pesticides and Their Relationship to Environmental Health. But these scientists were given relatively few public resources with which to study these problems. In 1958, for example, an advisory committee to the Department of the Interior recommended a research budget of $25 million per year to study possible threats to fish and

TABLE 8.3: U.S. Government Publications on Pesticides, by Topic

Year	Pesticide Support				Pesticide Hazards						
	Effectiveness	New Products, Other Support	Applying and Disposing	Total	Applicator	Consumer	Environment, Nonhuman	Measurement	Total	Alternatives	Foreign, Technical
1948	3	0	0	3	0	0	1	1	2	0	1
1949	9	3	0	12	0	0	1	1	2	0	6
1950	9	0	0	9	1	0	1	1	3	1	6
1951	9	8	0	17	0	0	1	1	2	0	3
1952	10	2	0	12	0	1	1	0	2	0	8
1953	9	0	0	9	0	1	0	0	1	1	9
1954	6	0	0	6	0	0	1	0	1	0	7
1955	6	0	0	6	0	0	2	0	2	0	2
1956	7	0	0	7	0	0	3	1	4	1	6
1957	5	0	0	5	2	0	1	0	3	0	8
1958	7	0	0	7	0	0	1	0	1	1	7
1959	7	1	0	8	0	1	1	0	2	0	9
1960	3	1	0	4	2	2	0	0	4	0	11
1961	4	0	0	4	0	2	0	0	2	1	9
1962	3	4	0	7	2	0	2	0	4	0	9
1963	2	1	0	3	3	4	2	0	9	1	7
1964	2	0	0	2	2	6	3	0	11	0	7
1965	6	0	0	6	3	2	3	0	8	0	13
1966	3	1	0	4	2	1	3	2	8	0	6
1967	7	0	0	7	1	3	2	3	9	0	15
1968	9	3	0	12	6	1	6	0	13	0	15
1969	8	0	0	8	1	4	4	0	9	0	17
1970	3	0	0	3	1	1	7	0	9	0	10
1971	3	0	0	3	1	0	3	3	7	0	7
1972	3	0	1	4	0	1	6	4	11	3	7
1973	1	0	0	1	2	2	12	3	19	1	6
1974	2	0	0	2	0	1	1	1	3	1	6
1975	2	1	2	5	4	2	11	3	20	2	18
1976	7	0	4	11	5	3	12	2	22	0	19
1977	7	1	6	14	5	4	7	1	17	0	8
1978	7	1	0	8	0	3	2	0	5	1	8

TABLE 8.4: Research on Pesticide Hazards

Year	Effect on Man	Carcinogenicity	Effect on Other Nontarget Species	Environmental Contamination	Interactions	Mutagenicity	Teratogenicity	Total
Before 1950	38	3	4	1	7	15	1	69
1950	5	2	5	0	2	1	1	16
1951	11	11	11	1	2	2	1	39
1952	3	5	4	0	1	2	1	16
1953	13	3	2	0	0	4	0	22
1954	1	2	1	0	2	4	0	10
1955	0	7	1	0	1	2	0	11
1956	13	5	2	1	1	0	1	23
1957	7	10	1	0	5	4	0	27
1958	13	3	2	2	4	2	0	26
1959	14	8	0	1	4	5	1	33
1960	16	8	7	1	5	6	0	43
1961	26	5	3	4	4	4	1	47
1962	23	6	4	1	6	10	0	50
1963	42	2	11	10	8	7	0	80
1964	55	8	4	7	8	5	4	91
1965	86	12	1	7	10	13	6	135
1966	44	7	0	19	12	19	2	103
1967	73	9	0	10	24	15	7	138
1968	69	2	0	22	20	21	11	145
1969	70	8	1	14	25	17	15	150

NOTE: Table represents citations listed by chapter, in the *Report of the Secretary's Commission on Pesticides and Their Relationship to Environmental Health,* (U.S. Department of Health, Education, and Welfare, 1969).

wildlife resulting from the growing use of pesticides.[17] The department and Congress recognized the need, but authorized only $230,000.[18] The *Congressional Quarterly* reported in 1965 that the Food and Drug Administration, responsible for protecting humans against the consumption of harmful chemicals, "has always been understaffed and underfinanced, in part because of fears in some quarters that the agency might be 'overzealous' in its regulatory activities."[19] In testimony in 1978 before Congressman Whitten's appropriations subcommittee, the director of FDA's Bureau of Foods said that the budget for the research necessary for food regulation was a victim of "systematic starvation."[20]

Published research on carcinogenicity included a few studies reported in the 1950s, but the bulk of those on which the HEW secretary's commission relied were published in a later period, 1965 to 1969. This later research provided the basis for the following commission findings, which became relevant in subsequent regulatory decisions: (1) the presence of carcinogenic substances in food may be a significant factor in

TABLE 8.5: Research on Alternatives to Pesticides

| | New Federal Research Institutions[a] | |
Year	Plan	Construction
1958		
1959		
1960	2	2
1961	2	2
1962	1	1
1963	1	1
1964	0	0
1965	6	1
1966		4
1967		0
1968		0
1969		1
1970		
1971		
1972		
1973		
1974		
1975		
1976		
1977		

[a] USDA-Agricultural Research Service. New installations to study insect control.

the occurrence of "spontaneous cancer"; and (2) since the effects of carcinogens on target tissues appear irreversible, reduction of exposure "may be one of the most effective measures toward cancer prevention." This 1969 panel recommended eliminating the use of DDT in food production.[21] Among the multiple health and environmental criteria on the basis of which other major pesticides were banned during the 1970s, evidence of carcinogenicity was always prominent.

While research findings on pesticide hazards increased during the 1970s, there was from 1949 through 1978 a continuing strong program of research on pesticide effectiveness. The pest control budget of the federal Agricultural Research Agency shifted toward basic insect research and nonchemical alternatives to pesticides,[22] with several new facilities for such research being constructed during the 1960s (see table 8.5), but state research and extension agencies were expected to continue their high priority on chemical control of local pest populations. "Rachel Carson did us a favor," said one research administrator. "Her book, *Silent Spring*, began the golden era in which we entomologists could simply put out our hands and the money was given to us."

While public research on chemical effectiveness and alternative pest controls is conducted largely by agricultural scientists, much of the significant research on pesticide hazards has been conducted by scientists in disciplines not closely associated with agricultural sciences. Scrutiny of the many citations in the 1969 HEW report reveals only a few from agricultural science journals.[23] As of 1964, according to one analysis, *all* USDA pest research funding was for improved pest control, with none at all for environmental monitoring or for the study of toxicology in nontarget organisms or in humans.[24]

Within agricultural research institutions, the readiness to stress the need for pesticides and to question claims of pesticide hazards is evidenced by the activities of an organization of agricultural scientists called the Council of Agricultural Science and Technology (CAST). CAST was formed to offer "objective" and "scientific" interpretations of controversial scientific subjects, such as those relating to pesticide hazards.[25] More than two dozen scientific associations belong to CAST,

166

but these are largely from disciplines committed to producing and integrating agricultural technology, rather than from the disciplines which study pesticide risks and other externalities. Within CAST's affiliate societies, it should be noted, there are individuals who have protested membership in CAST on the ground that it is an advocate organization.[26] Thus, the paradox of technical information being mobilized for purposes of manipulation, while successful manipulative efforts like Rachel Carson's influence research policy, thereby restructuring the pool of technical knowledge.

COMMENTARY

The groups, agencies, and concerned publics in the ex/al coalition have filled a void of research on hazards or side effects of agricultural technology. In pressing for regulatory laws, they have provided a measure of protection to humans and the environment. Most of this policy output has been achieved without the blessing or participation of the agricultural establishment. Indirectly, however, agricultural research institutions have felt pressure to seek alternatives to some existing technology, and also to pay more attention to side effects.

The two coalitions—the ex/al coalition and the industry coalition—have sometimes found it expedient to work together: for example, in support of research on alternatives to pesticides. Both in Congress and in the executive branch, agencies have worked to unite these coalitions. In July of 1977 it was announced, to the delight of environmentalists, that it would henceforth be a goal of the USDA to reduce reliance upon chemicals by increasing research on strategies that use a combination of biological, chemical, and other controls.[27] Secretary of Agriculture Robert Bergland's serious commitment to the development of new pest control strategies was manifested in a subsequent USDA report on organic farming and in budget emphases as well. The 1980 budget, requesting $6 million for integrated pest management, virtually doubled the amount Congress had appropriated for the previous year.[28]

The large public which had become concerned about pesticides should have been reassured by the subsequent efforts of federal regulatory and research agencies. News coverage continued to be high, however, as indicated in table 8.2, suggesting that that public continued to be informed about and involved in pest control policies, as did the ex/al groups and agencies which had helped to generate pest control awareness during the decades since Rachel Carson.

CHAPTER NINE

A Basis
for Integration

Both the industry coalition and the ex/al coalition have offered valuable, if contrasting, scenarios for meeting world food needs. In the industry coalition's scenario, of course, agribusiness corporations have opportunities to grow, and agricultural research institutions are assigned developmental roles. In the ex/al coalition's scenario, external goals such as preservation of the ecosystem have first priority, and the heroic roles are reserved for versatile, hard-working organic farmers. Each program offers perspectives from which the world's production systems are being shaped. Research agendas are implied in both, and from the two a balanced research agenda can be synthesized.

Agricultural economists usually present the industry scenario, and most agricultural scientists still prefer it. Major exponents of the ex/al scenario are biologists, like Rachel Carson and Barry Commoner, or humanists, whose academic credentials, if any, are in literature, philosophy, or history. Agricultural historians have taken the stage not as futurists but as realists, perhaps even as fatalists, whose work supplies evidence for exponents of both scenarios, while also rendering each a bit less fanciful than it might otherwise have been.

THE INDUSTRY SCENARIO

In the industry scenario, the agricultural industry saves the world. By increasing food production, it nullifies the doomsday prophecies of hunger, famine, and institutional collapse inspired by the specter of overpopulation. An innovative, production-oriented industry has shown it can carry the world's population through a period of transition from an earlier time when population was checked by disease and famine, to the emerging era, in which population will be stabilized by birth control. One recent and lucid presentation of the industry scenario is that of Australian agricultural economist Keith Campbell, in his book *Food for the Future.* [1]

Economist Campbell points out that world population has increased from 2 billion in 1930 to 4 billion at present, and can be expected to increase 50 percent by the year 2000. Fortunately, birth rates have begun to decline even in developing countries, and if trends continue, the world's population may stabilize at about 11 billion people. The task is to produce minimum diets for poor people while also meeting demands for luxurious diets from those who can purchase them. Presumably, world agriculture can meet this demand by adapting efficient production technology to sectors which do not now have it, by some expansion of croplands, and by the continuing improvement of our technology.

Admittedly, this technology is a heavy user of nonrenewable resources, especially of minerals and fossil energy in the form of fuel, fertilizers, and pesticides. But it is argued that such a technology may actually conserve soil by producing high yields on prime land and thus relieving the pressure to extend cropland onto hillsides and other fragile soil environments.

Campbell argues that fossil energy is, in the short run, relatively abundant: "It was estimated in 1974 that natural gas being wasted would be sufficient to produce . . . about three times that year's world consumption." [2] Crawford points to "huge reserves" of coal from which nitrogen fertilizer can be produced. Other fertilizer components—potash and phosphate—are abundant in the rocks of the earth's crust. [3]

170

Although water is also a scarce item in some environments, new irrigation technology is water conserving.

In short, the industrial scenario foresees expanded but prudent use of land, water, and other limited resources, and continues to look to the researchers for better conservation strategies. Campbell concludes: "The key to achieving the increased food output needed to feed the greatly enlarged world population in prospect, and to feed it more adequately, is a further application of science and technology to agriculture."[4]

His conclusion is shared by many distinguished students of the subject within agricultural economics.[5] For them, a major concern is, Who will do the research? Private industry undertakes little of such research, exponents of the industry scenario admit, because the benefits of basic and developmental research cannot be captured by its sponsors. Economist Ted Schultz concludes: "The only meaningful approach to modern agricultural research is to conceptualize most contributions as *public goods*."[6] Schultz and other analysts would like our decentralized public research establishment to undertake an international research agenda.

But public research, too, serves particular interests, as we have seen. The state budgets that largely fund experiment stations are expected to help local agriculture, and the promise of local and regional advantage brings political support to both state and national research institutions. If private industry lacks motivation to undertake an international research agenda, it does not follow that public research institutions, supported by the industry, are eager to devote themselves to international training and research goals.

A related difficulty, in the industry scenario, is in distributing food to those who need it most. Modern technology tends to be irrelevant to subsistence agriculture, or, indeed, tends to separate subsistence farmers from the land from which they drew sustenance and to convert its production to export crops. Critics of the industry scenario, including Lappé and Collins, have argued that developing nations could feed themselves adequately *without* the labor-saving technology produced in

171

western research institutions and imposed by multinational ag-ribusinesses.[7]

This distribution problem prompts three kinds of responses. One is to minimize the problem. Campbell disagrees with estimates that a large portion of the world's population now suffers malnutrition or hunger; he believes that subsistence farm families achieve better nutrition than is generally supposed, and that those released from agriculture can earn income to buy food.

A second response to the distribution problem is to define it away. Campbell says that those who cannot afford food do not, in an economic sense, demand it. "In the final analysis, the demand for food (as reflected by consumer purchases in market economies, and, to a lesser extent, the production activities of subsistence farmers) is a relevant criterion in any discussion about world food supply. In other words, it is what economists call 'the effective demand of consumers' that matters, not physiological needs as calculated by human nutritionists."[8]

A third response, direct and somewhat self-serving, has been to design compensatory programs to meet the food needs of the world's poor. A major program has been Title XII of the Foreign Assistance Act of 1974, passage of which was supported by the land-grant universities.

Title XII: Institution Building in Poor Countries

Agricultural scientists have long been prominent in training programs and technical assistance offered by U.S. universities for less developed countries (LDCs), despite the earlier, mistaken emphasis upon industrial development at the cost of agriculture. Efforts at agricultural training and assistance exemplify the dilemmas of cultural transfer. Students from developing countries enrolled in U.S. agricultural colleges have been taught labor-saving techniques by professors who seemed unaware that labor was a major resource in the countries from which their foreign students came. Moderate-zone agriculture has been taught to students from the tropics. U.S. training has produced specialists unsuited for countries which lack an infrastructure of other specialists. Professors have insisted upon pristine methodologies for plant breeding, even though it has been proven that the rough-and-ready methods are more successful in undeveloped country settings. U.S. agricultural science has been helpful primarily to capital-intensive agriculture (usually producing crops for export) in countries which have already achieved intermediate-level infrastructures, such as Brazil.

Title XII was an effort to help U.S. universities reach the poorest nations. It was also a step toward providing international financing for decentralized science—for placing the burden of international research and training upon donor and recipient governments while keeping decisions in the hands of the research institutions. The program was housed within the U.S. Agency for International Development, from whose budget the U.S. cost share would be provided, but it was run by a seven-member Board for International Food and Agricultural Development (BIFAD), the majority of whose members were representatives of agricultural research institutions. Congressional oversight was by the agriculture committees.

Title XII provided for four kinds of activities: strengthening the research systems of poor countries; providing support for several international research centers which had been established over the years by a consortium of governments and private foundations; financing research in U.S. institutions on food crops grown in the LDCs; and helping U.S. institutions to tackle the problems of developing countries.

As critics of Title XII had feared, the land-grant universities, and especially the agricultural colleges, found it easier to use the money to strengthen their on-campus capacity than to take the long step toward institutional development in poor countries. Title XII, if it has achieved little else, has symbolized a commitment by agricultural research institutions to the broad spectrum of agricultural development. At the very least, it has implied a recognition of the fact that different cultural settings may require alternative technologies.

It remains true that the industrial scenario seeks research for an agriculture characterized by large firms, heavy inputs of fossil energy, high mechanization, and low labor use, with products intended mainly for the people of developed countries. Yet the scenario concedes the increasing scarcity and high cost of major inputs—fossil energy, land, and water. It also accepts limits to environmental manipulation, and for these problems anticipates remedies from future basic research findings.

More research is *not* viewed as a remedy for the distribution problem, which is therefore passed on to the political realm for solution. In 1972 agricultural economist Willard Cochrane predicted that civil wars would be fought over the distribution of benefits from agricultural development.[9] He may be wrong: political revolution may be impossible once modern agricultural technology is in place.[10] Yet drafters of the scenario might find it hard to accept the likely political solutions: in Taiwan distributive solutions were effectively imposed by Japanese and Kuomintang dictatorships;[11] elsewhere, as in China and Cuba, the problem has been "solved" by socialist revolutions, in which the sorts of interests which support the industrial scenario would play peripheral and quite dependent roles.

THE EXTERNALISTS/ALTERNATIVES SCENARIO

In the scenario favorable to alternative agriculture, agriculture is invited to play a central role in man's effort to become reconciled with the environment in which he is obliged to live. Agriculture as the link between man and ecosystem has been the theme of Dr. Barry Commoner, an exponent of the ex/al scenario. Commoner points out that agriculture is part of two systems—the life-sustaining ecological system (ecosystem) and the system which produces man's goods. Agriculture must be constrained and shaped by both systems.

Commoner has stressed that agriculture plays a vital role in trapping abundant solar energy. Farmers need not be burdened by the increasing costs of fuel, electricity, and chemicals if they diversify their farming to include crops which can fix nitrogen in the soil, using animals whose manure is a necessary part of the energy regenerating process. For

Commoner, agriculture's crucial balance sheet should counterpose energy gleaned from the farm against energy which must be purchased and which is often inefficient.[12]

One embodiment of Commoner's efficient farm is the "organic" farm, which minimizes the use of synthetic chemicals. The organic farming movement in the U.S. has among its components organized farmers, publishing institutions, and marketing organizations. It has an ideology and a way of life; past and current prophets; and a large number of "true believers." Garth Youngberg, who studied the organic movement, wrote that much of its ideology could be summed up in the words of a well-known TV commercial: "You can't fool Mother Nature." He went on to observe: "Alternative agriculturists believe that modern man has lost touch with nature, that he has become insensitive to nature's intricate, delicate, mysterious, but, none the less, infinitely wise and immutable laws."[13]

One of the prophets of organic farming believers was the late British economist, E. F. Schumacher, whose book *Small Is Beautiful* has become a major document not only for organic farmers but for others who question the brashness of agricultural scientists who busily erect their beaver dams against the forces of nature, heedless of whether the disequilibriums on which they thrive will become impossible to maintain. Schumacher grieved that man no longer sees himself "as a part of nature but as an outside force destined to dominate and conquer it." Schumacher added that man "even talks of the battle with nature, forgetting that, if he won the battle, he would find himself on the losing side."[14]

There is particular emphasis in organic farming on the integrity of the soil and its central role in preserving life. "All food supplies in the world—even the fish in the sea—come directly or indirectly from the soil, chiefly from those few inches of earth known as top soil," says Charles Walters, another prophet of organic farming and editor of a major journal, *Acres*.

As Youngberg noted: "Organic farming is a multi-faceted concept reflecting a number of separate but interrelated ideological goals. It is the combination of these goals which gives the organic concept such a powerful ideological base." It is "an eco-agriculture movement." And it is this set of broad concerns, rather than any economic advantage, which causes organic farmers to undertake the laborious routines of "composting"—assembling of manure, leaves, and lime, which in rotted condition are returned to the soil to refurbish its nutrients and its body. The rewards of practitioners are expressed by one of them: "When I learned I could farm without chemicals, it made my soul feel good."[15]

Youngberg found, in interviews with organic farming leaders, a belief that the agricultural research establishment has "closed the books" on organic agriculture. Indeed, agricultural scientists tend to agree. Said one soil scientist: "It's all there, if someone wants to farm organically, the research is all there, it's all been done back in the forties and before."

The alternative agriculture scenario, though it seeks ecological and human values, still envisions a production process in which there is enough food for everybody. Frances Moore Lappé and Joseph Collins have argued that the world contains adequate food resources without the use of the synthetic fertilizers upon which the "Green Revolution" was based. They argue that even Bangladesh, a "basket-case" nation with its proliferating, malnourished population, can feed itself if its agriculture is fully developed and is devoted to production for local people rather than for export. Instead, argues the ex/al scenario, developing country agriculture is being converted away from plots producing for sale in local markets, toward large, capital-intensive farms which produce crops to be sold abroad and which may not yield as much value per acre despite large fertilizer inputs.[16]

The ex/al scenario offers an alternative world in which most countries are self-sufficient in food production and do not have to import expensive fertilizer and other inputs. The land and other natural resources are thus preserved rather than "mined."

A major problem for alternative agriculture is its assumption that many people, given an enlightened choice among career alternatives, will be willing to spend a lifetime of labor in agriculture—indeed, will gain pleasure and status from such labor—and, that consumers with increasing incomes will be willing to accept simpler diets, thereby reducing the demand for the imported foods which are now grown on fertile lands in developing countries.

Such expectations are contrary to the trends in all mass societies. Ex/al writer Wendell Berry grants that it will be difficult to manage "the vulnerabilities of the human personality", which usually would choose mechanical power and luxury food. Pessimism creeps in in any honest confrontation with the question "Can we, believing in 'the effectiveness of power,' see 'the disproportionately greater effectiveness of abstaining from its use'?"[17] Berry admits that the only group he knows who have been willing adequately to restrain themselves in energy use are the Amish.

Ex/al groups should therefore welcome techniques and processes which minimize the need for sacrifice and inconvenience. In this regard, alternative agriculture has a big research agenda, much of it in agreement with the agenda of the industry scenario. For example, both scenarios would benefit from the breeding of plants which are capable of fixing their own nutrients and are more naturally resistant to threats from the environment. Both scenarios may benefit from a better understanding of nutrition and growth in humans as well as in animals. In both scenarios, there must be techniques for conserving soil and for efficient storage of food.

Such mutual objectives are among the major concerns of the research agenda suggested by the 1975 World Food and Nutrition Study, summarized in table 9.1. The unique objective of this study was to suggest a research agenda which would serve the food needs of the world's poor people; and by focusing on world hunger rather than industry interests, the WFNS produced a kind of ex/al agenda. Another agenda was created at a Kansas City conference, also in 1975, at which most delegates were supporters of the industry scenario.[18] It turned out that the agendas were in agreement on many objectives. The Kansas City conference report produced a set of "most important priorities," agreed upon both by delegates and by panels of agricultural scientists. As table 9.1 indicates, many of these specific proposals can be subsumed under the more general suggestions of the World Food and

A Basis for Integration

TABLE 9.1: A Research Agenda to Meet World Food Needs

Recommended Research Priorities, World Food and Nutrition Study[a]	Number of Specific Priority Recommendations, Kansas City Conference[b]
Nutrition:	
Determine damage caused by various kinds and levels of malnutrition: effects of diet patterns on levels of human functioning.	6
Improve effects of direct intervention programs: evaluate effectiveness of alternative programs in reaching nutritional goals.	21
Determine specific foods that best meet nutritional needs under differing circumstances: effects of individual nutrient levels, as consumed, on nutritional status.	24
Improve effects of full-range government policies: effects on nutrition of policies and practices usually formulated with no consideration of possible nutritional consequences.	8
Food Production:	
Plant breeding and genetic manipulation—strengthen tools of genetic manipulation: plant breeding and classical genetics; cell biology; genetic stocks.	75
Biological nitrogen fixation—increase biological nitrogen fixation associated with major crops: improve recognized symbiotic associations; attempt to establish N2-fixing associations with grains and other nonlegumes; transfer fixation capability from bacteria to plants.	12
Photosynthesis—increase amount of photosynthesis in major crops: reduce photorespiration and dark respiration; transfer traits from photosynthetically efficient plants.	8
Resistance to environmental stresses—improve resistance of major crops to drought, temperature extremes, deficiencies of acid soils, salinity: rapid screening techniques; tolerance of acid soils; shorter season crops; larger root systems; better use of soil fungi; salinity resistance; farming systems.	38
Pest management—reduce preharvest losses due to pests: integrated pest management; specific control mechanisms.	18
Weather and climate—improve techniques for predicting weather and climate and using information to assist adaptation by farmers: reduce weather damage to food production.	12

178

TABLE 9.1 continued:

Recommended Research Priorities, World Food and Nutrition Study[a]	Number of Specific Priority Recommendations, Kansas City Conference[b]	
Food Production con't.:		
Management of tropical and other soils—improve management of tropical soils to increase crop productivity: soil classification; land-clearing methods; correcting soil deficiencies; maintaining desired soil characteristics; suitable cropping systems and technologies.	11	
Irrigation and Water Management—improve management of water supplies: adjustment in farming systems and management of water movement for optimal supply to crops; adapting farming operations to water availability.	2	
Ruminant livestock and other animals—increase product yields from ruminant livestock, particularly in the tropics; forages; priority animal diseases; genetics and reproduction.	(Beef	10
	Lamb	7
	Goats & Rabbits	9
	Dairy	6
	Pork	8
	Poultry	5
	Total	45)
Aquatic food sources — increase contribution of aquatic resources to world food supply: waste reduction and upgrading product through processing; aquaculture research in both breeding and seed supply and polyculture management.	10	
Farm Production Systems—improve production systems, particularly for small farms in developing countries; methodology for identifying appropriate farming systems; multiple cropping; soil and water management; equipment-labor relationships.	50	
Food Marketing:		
Market expansion and new food sources—extend market scope for consumers and farmers in developing countries; enhancing purchasing power; transportation; marketing institutions; managing marketing flows of major commodities.	31	
Postharvest losses and food technology—reduce postharvest losses: nature and magnitude of losses; pest control after harvest; food preservation for humid and arid tropics.	45	

179

TABLE 9.1 continued:

Recommended Research Priorities, World Food and Nutrition Study[a]	Number of Specific Priority Recommendations, Kansas City Conference[b]
Food Marketing con't.:	
National food policies and organizations—improve policies and organizations affecting food production, distribution, and nutrition in developing countries: human performance in food systems; comparative studies to identify transferable improvement factors (decentralization, local participation, staff development); interactions of income distribution with food production and nutrition; methodology of sector analysis.	11
Policies and Organizations:	
Food reserves—improve role of reserves in relation to other measures for stabilizing food supplies: improving developing country food reserve practices; identifying improved mixes of reserves and other measures to stabilize food supply.	4
Information systems and analysis of rural life—improve flows of information in support of decision making on food and nutrition: producer information needs to use better technology; crop monitoring systems; international data bases on land uses and malnutrition; information systems design.	12
Trade policy and finance—improve effects of trade policy on food production and nutrition; studies on effects of trade liberalization; consequences of international management of trade; optimum trading patterns.	14

[a] Table 3, "Recommended Research Priorities," in *World Food and Nutrition Study: The Potential Contributions of Research* (National Academy of Sciences, 1977).
[b] The number listed is the total number of "most important problem statements" so designated both by conference delegates and by panels of agricultural scientists at the Kansas City conference.

Nutrition Study. At the Kansas City conference, for example, seventy-five of the "most important" recommendations dealt with plant breeding and genetic manipulation.

While this research agenda may serve both scenarios, it apparently does not necessarily serve the interests of the politicians and administrators who control the research process. For example, the chief funding mechanism suggested for implementing this agenda, recommended by virtually all review panels, is the competitive grant, which encourages innovation, maintains research quality, and also taps a wide pool of scientists. But institutional administrators—and Congressman Whitten—have been unenthusiastic about any agenda offered under competitive grants. At the 1980 hearings Congressman Whitten expressed his reservations about such basic research: "With all the need that we have, how are you going to shift it over just to find some curious answers in the hope that you just might satisfy your curiosity at some time in the future?"[19] Because of such opposition, paradoxically, the agenda which serves both scenarios probably cannot be implemented without strong support from within both coalitions.

Beyond the shared agenda, of course, there remain subjects upon which there is strong disagreement. The industry coalition wants to calibrate high technology packages of chemicals, machinery, and genes—for example, machines that displace labor—and in seeking these objectives they may expect continuing opposition from the ex/al coalition. To escape the political effects of this opposition, industry groups have already begun to seek funding through taxes on their own commodities, earmarked for research and promotion, rather than through direct support from state or national appropriations. Meanwhile, the ex/al coalition appropriately seeks out nonagricultural research institutions to perform research that would expose the health hazards generated by agricultural chemicals. There will continue to be a need to reconcile research and other policy outputs which each coalition produces within its own institutional domain.

Thus, those who would integrate the two research agendas have two tasks: first, gaining implementation of an agenda that is justified under both scenarios; and, second, reconciling the sometimes conflicting individual agendas. We turn now to an examination of agencies which have shown a capacity for integrating research policies.

181

CHAPTER TEN

Integrating
Agencies

By 1980, erstwhile critics of agricultural research institutions were beginning to incline warily toward supporting them, even as industry support for agricultural research was being rekindled. Some credit for this movement should be given to research administrators and scientists, although some difficulties in developing support must be laid on them also, since the judgment persists that too many research leaders are narrow, peevish, and unaggressive, while too many scientists are unimaginative, mediocre, and unproductive. Of course, the institutions should also be judged by their output, which is excellent, according to cost effectiveness studies. But the new visibility and stature of agricultural research, and indeed its new growth posture, is less the research establishment's doing than that of the congressional agricultural committees and the secretary of agriculture, among others whom we shall refer to in this chapter as "integrating agencies." It is the job of these political centers to mesh demands from many sides—in this case, to reconcile the demands of the industry with those of its critics.

Other integrating agencies have sought to implement the reconciled agenda, including the late Science and Education Administration, which was the USDA's most recent effort to coordinate all agricultural research. Agencies which aggregate new demands may not always seem to be supportive of those who had sought to meet earlier demands, particularly in the eyes of agencies which were previously sustained by a specific and rather narrow clientele. One integrating agency, the president's Office of Management and Budget, seems anything but helpful in its insistence on "accountability," in its antipathy to the pork barrel process on which research has depended, and in its impulse to remove budget dollars which are not bolted down by legislative mandate, such as those for research. Even OMB, however, may promote growth. Scrutiny by a neutral agency may symbolically reassure concerned publics that the research institutions are performing adequately.

To add one more to the list of agencies to be discussed in this chapter, one can assert that the international research centers, which were created by private foundations and which have been supported in considerable part by the U.S. Agency for International Development, also have an integrating function in adapting our technology to the environments of developing countries.

THE HOUSE COMMITTEE ON AGRICULTURE

As late as the 1960s, the power of the House Agriculture Committee was wielded by a few senior members in both parties who were, individually, spokesmen for one or two major commodity interests, particularly for cotton, tobacco, rice, peanuts, wheat, and feed grains.[1] These senior members were also quite conservative, as indicated by *Congressional Quarterly* Conservative Coalition scoring, and their committee served as a vehicle for industry interests (illustrated in the passage of the 1947 pesticides law, discussed earlier).

Lately, the seniority system, which had vested power in the southern, Plains, and midwestern commodity representatives on the committee, has been weakened. An influx of liberals unseated the chairman, Robert Poage (D., Tex.), in 1975, replacing him with Congressman Tom Foley (D., Wash.), who was from a wheat district but who had an orientation toward broader concerns. Subcommittees were created for some noncommodity concerns, including rural development and research, several of whose members were instrumental in merging in-

TABLE 10.1: Characteristics of Districts of Members of House Agriculture Subcommittee on Department Investigations, Oversight, and Research

Majority Party	Research Installations & Experiment Stations
De La Garza (TX)	Screwworm Research Lab, Mission
	Subtropical Area, Weslaco
	Fruit, Veg., S&W Res., Weslaco
	Cotton Insects (Brownsville)
Brown (CA)	Area Office, Riverside
	Salinity Lab, Riverside
	Citrus Research, Riverside
	(UC —Riverside)
Skelton (MO)	None
Breckinridge (KY)	(UK—Lexington)
	(State U—Frankfort)
Hightower (TX)	Guar Res., Vernon
	Gt. Plains Res., Bushland
English (OK)	S. W. Livestock, F. Res., El Reno
	S. W. Ct. Plains F. S., Woodward
	Hydraul. Lab, Stillwater
	Crop Genetics, Stillwater
	Hard Red Winter, Stillwater
	(Oklahoma State)
Fithian (IN)	Insect Control, Purdue
	Water Erosion, Purdue
	Poultry Genetics, Purdue
	Plant Science, Purdue
	(Purdue)

dustry and ex/al interests within the 1977 research law (Title XIV, Food and Agriculture Act of 1977). The House committee became the workplace of liberal activists whose major concerns included solar energy, civil and human rights, small farms, and in the case of Congressman Fred Richmond (D., N.Y.), urban poverty as well. Richmond and Congressman George Brown (D., Calif.) were major spokesmen for the ex/al research agenda. Meanwhile, the average of Conservative Coalition voting scores for Democratic members dropped from 64 percent in 1959 to 44 percent in 1976, and for Republican members,

TABLE 10.1 continued:

Majority Party	Research Installations & Experiment Stations
Jenrette (SC)	Cattle Breeding, Pee Dee
	Cotton Insects, Pee Dee
	Coastal Plains S&W, Florence
Thornton (AR)	(AU—Pine Bluffs)
	(AU—Monticello)
Ammerman (PA)	Reg. Pasture Res., University Park
	(Penn State)
Minority Party	
Thone (NE)	Soil, Water, Animal Waste, Lincoln
	Plant Sciences and Entomology, Lincoln
	(Nebraska University)
Heckler (MA)	None
Grassley (IA)	None
Marlenee (MT)	U.S. Range Livestock Exper. Stat., Miles City
	N. Plains, S&W, Sidney

from 94 percent to 74 percent. In the turnover after the 1980 elections, replacements in both parties were mainly conservative, but Congressman Brown has became chairman of the research subcommittee, and a few other liberals have moved into leading positions. It would appear, from the character of the House committee's recent membership and leadership and from its integrative role in the passage of the 1977 research law, that the House committee is well prepared to represent both ex/al demands and the demands of commodity interests. In addition, several subcommittee members have a district interest in research facilities (table 10.1).

THE SENATE COMMITTEE ON
AGRICULTURE, NUTRITION, FORESTRY

The Senate Agriculture Committee also has a subcommittee for "agricultural research and general legislation," which has so far not exercised significant influence upon research policy, through its membership or its staff. In recent years the full committee's main impact upon agricultural research policy derived from its serving as a cockpit for two activist liberals, the late Hubert Humphrey (D., Minn.) and former member George McGovern (D., S. Dak.). The committee is on the whole still populated with conservatives, as in earlier years, and indeed the average of Conservative Coalition scores for 1981 is higher than for 1959 (for the Democrats, up from 46 percent to 53 percent; for the Republicans, from 79 percent to 83 percent).

Humphrey was a principal author of the 1977 research statute and the 1972 Rural Development Act, as well as Title XII of the Foreign Assistance Act of 1974, which offered land-grant universities a key technical assistance role. Sen. McGovern led in establishing a Select Committee on Nutrition and Human Needs, whose membership was drawn equally from the Senate Labor Committee and the Senate Agriculture Committee. This select committee helped develop the food stamp program, and later called attention to human nutrition as an objective in food aid and agricultural research programs. McGovern and Humphey were also leaders in developing Food-for-Peace, under which a large overseas research program was implemented.

In these activities, Humphrey and McGovern were helping to create a public that was concerned about nutrition at home and abroad. This concerned public is to be found both in the agricultural states from which these senators came and in the large urban states where both senators sought votes in presidential races. All Senate members must now deal with this concerned public. In the near future, however, the Senate committee will likely be responsive to industry interests, for, like the House committee, it has members attentive to major commodity interests within their states. In addition, in the 1981 turnover of Senate members the committee gained a conservative chairman, Sen. Jesse Helms (R., N.C.).

Both the Senate and House congressional agriculture committees, having created the 1977 research charter, are also in a position in the budgetary process to recommend funding levels. In future legislation, however, the legislators may even mandate funding levels according to formula in order to strip budget decision makers of the power to ignore the directions formulated in new research funding authority, as they did in the case of both the Rural Development Act of 1972 and the 1977 research statute (Title XIV of the 1977 Food and Agriculture Act). It remains to be seen whether the agriculture committees, as the new monitors of agricultural research, will continue to serve as integrating agencies.

USDA—"LEAD AGENCY"
FOR AGRICULTURAL RESEARCH

The secretary of agriculture sits atop a distinguished but changing bureaucracy. Historians Rasmussen and Baker, surveying the Department of Agriculture's history from 1862 to 1962, found an "old" and a "new" department; and one can argue that there have now been two additional eras or departments: the post–World II period, in which the USDA was in service to an industry clientele; and the present period, in which it is responsive also to interests present in the ex/al coalition.

"old" USDA, from its founding in 1862 to 1932, developed as an excellent educational, scientific, and statistical agency.[2] A "new" department evolved during Secretary of Agriculture Henry A. Wallace's administration, in which the earlier functions of science and education shared importance with a host of action programs, including commodity price supports and market development, and also with major programs for soil conservation, rural electrification, farm worker relief and resettlement, and food assistance to needy families. With these new functions, the size of the USDA increased from twenty-two thousand full-time employees in 1932 to seventy-nine thousand in 1948, which is approximately its current level.

After Wallace, this broad-ranging department entered a third stage, in which its major programs were run by and for farmer clients. Programs which had been constructed to permit grass-roots decision making through local farmer committees invited client dominance. By the 1960s, Theodore Lowi counted "at least ten separate autonomous local self-governing systems."[3]

The USDA, under governance by the farm organizations and legislators within the industry coalition, phased out programs for farm workers, rural communities, and food consumers. One secretary of agriculture, Charles Brannan, made an unsuccessful effort to restore the goals of human nutrition and equity for small farmers, and in the 1960s Secretary Orville Freeman announced some nonindustry goals such as "rural renewal." But the respective presidents under whom Brannan and Freeman served felt obliged to release the USDA to its industrial clientele, as the price for obtaining cooperation from influential senior congressmen. "Placate them," President Kennedy told Secretary Freeman, referring to Congressman Whitten and to two other southerners who chaired the House and Senate agriculture committees. As a result of this presidential strategy, Freeman's administration was unable even to support rapid expansion of food stamps.[4]

During these years, Republican Secretary of Agriculture Ezra Benson (1953–60) made efforts, successfully resisted within Congress, to reduce commodity subsidies, and secretaries Clifford Hardin and Earl Butz (1969–74) tried to transfer out of the USDA the burgeoning food assistance programs.

We may postulate a fourth stage in the USDA's history, one which began with an announcement during the Butz administration, by Assistant Secretary Don Paarlberg, that the externalities groups had imposed a new agenda upon the USDA, including such objectives as low food prices, food assistance, environmental protection, rural development, protection of civil rights, and collective bargaining for farm workers.[5] This agenda was addressed by Secretary of Agriculture Robert Bergland, himself a farmer, who chose his assistant secretaries from among representatives of the ex/al coalition. These included Carol Foreman, formerly the director of the Consumer Federation of America; Rupert Cutler, a former editor for the National Wildlife Federation; Alex Mercure, a grass-roots leader among rural Mexican-American people; and Howard Hjort, who had gained a reputation as a "populist" economist.

These assistant secretaries maintained an open-door policy toward public interest groups. They made speeches and statements proclaiming new directions for the USDA. For example, a speech by the assistant secretary for research and conservation, Rupert Cutler, calling for reduction in the use of agricultural chemicals in order to protect the environment, was hailed by the Environmental Defense Fund as "a basic shift in policy." Maureen Henkel of EDF is quoted as saying: "The change is so remarkable I can't believe it. I have to hold myself down from walking two feet off the ground." The effect upon senior USDA officials who heard Cutler's speech was also striking. According to one observer: "They sat there just looking at him; their faces were white, their mouths dropped, and their eyeballs rolled back in their heads." One regional administrator commented, "He'll finish us yet."[6]

The deputy assistant secretary for research, James Neilsen, speaking to senior administrators of the Agricultural Research Service, put forward eight priorities of the new administration, among which the traditional mission of "productivity" was conspicuously listed last. The eight priorities were: human nutrition research and education, an effective prelude to a more meaningful and appropriate farm policy; energy conservation; land and water conservation and management; pest management; environmental protection; service to special disadvantaged groups, especially minorities and women; production, and production efficiency.[7]

The Bergland administration was quick to challenge any effort to transfer food, welfare, or developmental functions from the USDA to other departments. The USDA aggressively sought and obtained "lead agency" status for nutrition research, though it had admittedly neglected this subject previously.

The USDA's open-door policy and its strong statements of new directions earned immediate support from most consumer and conservation organizations. Said one consumer advocate: "We have better access to the USDA than any other agency, and it [the USDA] has moved quickly from the nineteenth century to the most forward-looking agency in government."

Agricultural research leaders were disappointed that the adminis-

189

tration was more directive than it was supportive, but one could argue that Bergland's administration did support research institutions, by redirecting them. Exciting new missions could attract new support and could save the agency from the extinction which Congress is said to visit upon agencies hopelessly caught in a clientele web.

Yet the industry coalition viewed the actions and attitudes of the Bergland administration as counterproductive, on balance. Research agency administrators found it difficult to consult with their "boss," Assistant Secretary Cutler. Cutler's public statements stressed things to be done, barely touching upon agency accomplishments. Cutler was regarded as an extremely weak participant at Bergland's subcabinet meetings, in which the assistant secretaries competed for funding and authority. Some industry clienteles became victims of research cuts: food processors would be expected henceforth to do most of their own research; the need for research on labor-saving machinery was called into question; tobacco research was terminated on the grounds that it was inconsistent with public health goals. A "tobacco" legislator, Congressman William Natcher (D., Ky.), conveyed to Secretary Bergland the anger of a displaced industry clientele: "I think you are wrong, Mr. Secretary, and I think you are running with the hounds. I say that to you frankly, I think you are running with the hounds."[8]

Although Bergland's administration and the research agencies wasted no love upon one another, they worked more or less in harmony to develop an integrated research agenda. T. W. Edminister, the administrator of the Agricultural Research Service, adequately supported the administration's request for human nutrition research in the face of repeated tongue-lashings from Chairman Whitten.[9] The Bergland administration, in reorganizing the agricultural science establishment, clearly intended to strengthen these institutions by improving scientific training, improving coordination and accountability, and forging positive links with users and with the broader science community. However, Bergland failed to provide adequate funding for the revitalization of the research institutions—less from lack of desire than from the "Catch 22's" of budget making.

Under the Reagan administration, another shift in emphasis took place. The administration supported increased funding for agricultural research and extension institutions, even as it slashed other agency budgets and reduced or abolished many other grants-in-aid to state and local governments. Secretary of Agriculture John Block and his associates, as advocates of the industry scenario, emphasized the need for productivity research to meet increasing food and export demands. There was no immediate retreat from the USDA's renewed concern for soil and energy conservation, and no immediate repudiation of research thrusts seeking alternative technologies such as integrated pest management and organic farming. Under Secretary Block, however, the Science and Education Administration was eliminated. This decision was justified on grounds of improving efficiency: the number of administrative jobs would be reduced and decision making simplified. Yet SEA had been conceived as a means of broadening the perspective of research administration and also as a way of mobilizing research in pursuit of national goals. In terminating it, the administration affirmed the autonomy of the state experiment stations and extension services. This and other USDA reorganizations under Secretary Block were intended to reorient the department toward industry interests.

THE SCIENCE & EDUCATION ADMINISTRATION (SEA) IN ACTION

Although the SEA no longer exists as an agency of government, its efforts at synthesizing an agenda acceptable to both industry interests and ex/al groups produced some significant results. Moreover, because its demise provides a clear example of the difficulties inherent in such an undertaking, a brief look at its history is relevant here.

After Congress had designated the USDA as "lead agency" for food and agricultural research, Secretary Bergland's administration created the Science and Education Administration to implement this mandate. The person selected to head SEA was Anson Bertrand, a respected soil scientist and research administrator at Texas Technical Institute (a non–land grant establishment), who won the confidence of both the administration and the research institutions.

SEA was given a long list of objectives: it was to coordinate several science agencies, to organize roles for key administrative leaders in the improvement of scientist training and recruitment, to develop award systems for employees, and to help in other ways to establish a productive work environment. Another rather explicit objective was to develop political support by means of adequate service: according to the first listed "criterion of success" in SEA's instructions, SEA was to be "visibly responsive to the public interest and needs in the food and agricultural sciences and thus tie programs to the needs of the support base."

The support base was identified, more or less, in the membership of a statutory Users Advisory Board. That board included representatives from the major segments of the agricultural industry (even farm workers!), from the federal action agencies which used agricultural research, and from consumerist and environmentalist groups. Soon after being organized, the user board's diverse membership from industry and ex/al coalitions achieved a compromise strategy of seeking increased research funding for all of the various members' major objectives: "The Board is alarmed by the declining support for agricultural research and development provided by the U.S. Government," its first report stated, noting that "such investments have yielded benefits to both producers in the form of lower costs and to consumers in the form of lower prices." The board strongly advocated a competitive grants program to stimulate innovative research and to involve nongovernmental scientists, but not "at the expense of either in-house USDA research or the established cooperative research apparatus." The board made the Carter administration its first target by asserting that the 1980 executive budget had not adequately supported institutional research. The board argued that SEA's administrator should be given a place in Bergland's departmental cabinet so that he could have direct influence upon administration decision making (this was done). Meanwhile the board looked for ways to generate public support for agricultural research.

SEA was also responsible for developing research priorities, and in this task it was to be aided both by the Users Advisory Board and by

192

another statutory group, the Joint Council, composed of representatives of research producers. The Joint Council represented not only the various institutions and functions within agricultural science but also nonagricultural science establishments. At a joint meeting between the Joint Council and the Users Advisory Board in 1979, these groups agreed upon a set of "items of high concern." This integrative statement of priorities included the following: redirection of programs to emphasize less energy-intensive production; processing and distribution practices which are also economic, helpful, and environmentally sound; increased focus on research to use alternative energy sources; increased focus on nutritional quality, safety, availability, and costs of food; increased efforts to expand basic and applied research, extension, and education programs in food and agriculture; a comprehensive review and analysis of the structure and organization of research, extension, and education; expansion of programs to improve social and economic standing of nontraditional clientele; and a review of the roles of Joint Council and Users Advisory Board with respect to international agricultural research.

The surprising harmony in stating priorities—and mutuality in seeking greater funding—were not characteristic of day-by-day efforts to develop joint planning and coordination among the agricultural science institutions. Conflict among research institutions was not unusual, particularly when efforts were being made to impose federal direction upon the state institutions. While attempts to coordinate agencies were largely unsuccessful, SEA's leadership explored research approaches that integrated the values of industry and ex/al coalitions, one of these being coordination of chemical and nonchemical means for pest control. Organic practices were also reintroduced as a research subject by a scientist panel quietly organized under the chief of SEA's planning staff, Ernest Corley. The panel's recommendation that more attention should be given to successful organic farming patterns was heralded as a major shift for agricultural research.

SEA's intent to institutionalize concerns about and to develop research on alternative technologies, as well as to achieve some coordination among science agencies, was a response to a prevailing mood among budget makers, as Don Paarlberg pointed out at the USDA's 1980 Agricultural Outlook Conference. Paarlberg noted, however, that efforts in these directions challenged both the traditional autonomy of the experiment stations and their traditional concentration on production efficiency, with the result that "the experiment stations, with their proud history, understandably resist this effort." Paarlberg argued, nevertheless, that SEA should exercise a coordinating role. "The Department of Agriculture is central, it is directly involved in the acquisition and distribution of Federal funds, and it comes closer to perceiving the broad public interest than does any other unit in the system."[10]

In another sense SEA offered leadership in reaffirming the partnership between state and federal research agencies, which had suffered greatly from the strain of recognizing new national priorities and from bickering over the division of shrinking research budgets. Emery Castle, in an address to the National Association of State Universities and Land-Grant Colleges in November 1980, noted that the partnership was in jeopardy: "The Food and Agriculture Act of 1977, the numerous constituencies that must be served by USDA, together with the multiple ties between higher education and the Federal Government, raise questions as to whether the partners still are marching to the same drummer. What happens during the next four years probably will decide whether the point of no return on the road to dissolution of the partnership has been passed."

Although SEA's administration was positive and conciliatory in thrust and its director, Anson Bertrand, was well regarded, it met with frustration and, in time, hostility. The planning staffs for state and federal research, though integrated on paper, were reluctant to cooperate in good faith. Bitterness resulted from procedures in which representatives of one research arm were expected to review the budgets of another. The national agencies which had been subsumed under SEA—the Office of Extension, the Cooperative States Research Service, and the Agricultural Research Service—may have felt, with reason, that they were losing both their staffs and their functions. There was resentment also over the fact that SEA's coordination staff was

194

given responsibility for organizing the budget. SEA's evaluation staff was perceived as being used against the researchers, and many agricultural scientists became hostile toward SEA.

SEA was politically vulnerable as a creation of the Bergland administration, which was displaced in 1981 during the presidential election and change of administration. The Reagan administration was persuaded to restore the individual status and previous identity of the SEA's component agencies, including the Agricultural Research Service and the Cooperative States Research Service. The reorganization was accomplished in June of 1981, in an order abolishing SEA. Appointment of an assistant secretary for research and education was promised.

THE OFFICE OF MANAGEMENT & BUDGET & THE WHITE HOUSE

Among the various integrating mechanisms, the federal budget is the most encompassing. In theory, the president's Office of Management and Budget (OMB), which prepares the budget, selects those programs which are cost-effective and which are attuned to the president's objectives. But in practice, pressures for economy at any price prevent the consistent use of general criteria in constructing the national budget. In the interest of balancing the total budget, OMB is under great pressure to reduce items such as research which are not already fixed by federal law or by political understandings. That pressure, according to one OMB official, "is indeed the principal explanation for agricultural research budget cuts."

But it should be noted that the savings are small from relatively tiny budgets such as that for agricultural research, and that by the same token there is little political impact from a doubling of research budgets. So it would seem that the president could enhance the research establishment on his own initiative if he saw the merits of doing so. Indeed, a large increase in funding for agricultural research is merited, according to the cost-efficiency criteria typically used in making budget decisions, because many studies have shown that returns on investments in agricultural research are in the range of 40 to 90 per-

195

cent. Budget officials who are aware of these studies, however, are inclined to portray them as self-serving, since most of them were produced within the agricultural research establishment.

OMB officials are well aware of the unfavorable evaluations of agricultural research—including the reports by the Pound committee and the General Accounting Office—which have found mediocrity, duplication, and lack of accountability. Most of these evaluators, of course, recommended both reforms and increases in funding.

The severely negative judgments on agricultural research within OMB appear to result less from evaluation of the research product than from lack of confidence in the process of research decision making. OMB people are sure that pork barrel decision making typifies agricultural research, and they are well prepared to cite flagrant recent cases. OMB was the effective advocate for shifting to a competitive grants system for funding increases in agricultural research, and was willing to provide large funding increases through that system. The appropriations committees, however, usually shifted most of the funds back into the pork barrel. Such pathologies in the budget process help explain the lack of growth in agricultural research budgets.

There is occasional attention from the White House to the needs and potential of agricultural research. Although research policy is not likely to be an important election issue, promises of adequate food, citizen health, and resource conservation, to be achieved in part through enhanced research programs, would seem to be acceptable within the programs of either national political party. Several such presidential initiatives in other areas have achieved public acclaim—among them President Kennedy's space program, President Nixon's cancer research plan, and President Carter's energy research effort.

Presidents and other leaders contribute to research policy by beckoning citizens to a future-oriented perspective, because the first step beyond aspiration is likely to be research. Indeed, research may be one case in which growth in budgets produces innovation. A larger executive budget, interestingly, could make unnecessary the pork barrel strategy which research agencies have used. In the presence of watchful groups, even increased institutional funding can lead to institutional change—and to institutions that are more, rather than less, responsive to the emerging era of limited resources.

THE INTERNATIONAL RESEARCH CENTERS

A set of international research centers has grown up—originally under sponsorship of the Rockefeller and Ford foundations and now financed by the United States and many other governments, as well as United Nations agencies—whose objective has been to adapt developed-country knowledge, germplasm, and research techniques to conditions in developing countries, mainly those in tropical zones. This effort began in the early 1940s with Rockefeller-sponsored research in Mexico to increase both the production and the nutritional content of wheat and other grains, the success of which earned the Nobel Prize for its leader, Dr. Norman Borlaug. Based on this experience, Rockefeller and Ford in 1960 jointly established, in the Philippines, the first center for the study of rice. By 1976 the number of international centers had grown to ten.[11]

The wheat and rice strains developed at these centers, which produced greatly increased yields, were largely responsible for the "green revolution." And although the social, political, and economic impacts of the green revolution have of course become controversial, the impact of the centers upon agricultural research has been integrative of the values discussed earlier. The centers have worked as closely as possible with the research institutions of developing countries to train researchers in techniques which do not require support from complex infrastructures, and they have developed a cadre of career scientists prepared to argue with their U.S. colleagues as to the merits of their distinctive techniques. They have also placed emphasis on crops and animals grown mainly for domestic consumption.

On the other hand, these centers, confronted with the challenge of the tropics, have also supported the merits of some advanced agricultural technology—in irrigation, chemical pest control, grain storage, and other areas. The international centers have become a third party offering useful commentary on the conflicting scenarios for future world food production.

197

CHAPTER ELEVEN

Summary & Conclusions

Agricultural research institutions have, in interaction with economic forces, cultural preferences, and other public policies, had a major influence upon our agricultural industry, and they, in turn, have been shaped chiefly by decisions of the industry subsystem. Inside this sub-system there has been agreement on some research goals, such as re-ducing human labor in agriculture and, in general, increasing agricul-tural productivity per unit of output (although some farm groups have blamed researchers when large supplies have caused low prices), while other potential research goals have been explicitly discouraged, among them those of preserving the physical environment, encouraging rural community development, and improving consumer nutrition. The subsystem has not been without conflicts, but these have arisen mainly over how research funding should be distributed, among commodities and among research agencies.

Lately, groups outside the subsystem have coalesced to demand that technology policy not be allowed to ignore unwanted consequences such as environmental contamination, risks to the health of consumers, and the wasting of natural resources, and they have advocated alterna-tive technologies which presumably avoid these unwanted side effects. These externalities / alternatives (ex/al) groups have questioned the wisdom of a technology that causes continual drastic reductions in the number of people in agricultural and rural America. Environmentalist and other ex/al groups have raised these large issues with extraordinary success in a political environment otherwise preoccupied with questions such as which town or congressional district should have a new research facility and which candidate will win next year's elections.

The raising of externalities issues coincided with the raising of questions by some science professionals as to whether agricultural research management is sufficiently innovative, efficient, and accountable. The Pound committee and other scientist evaluators found mediocrity and waste, presumably more than is usually tolerated with research organizations. At the same time, studies of research productivity invariably showed favorable cost-benefit ratios, but the notion prevailed that agricultural research could be improved through better management: it could be made more responsive to the full range of impacted groups, more attentive to large societal goals, less a prisoner of industry groups, and indeed even more proficient than it had already proven to be. The notion spurred new budget and planning processes, and also reorganizations, notably that which instituted a new USDA agency—the Science and Education Administration—overarching the Agricultural Research Service, state extension services, and the state experiment stations. The 1977 law which established SEA also provided for new research mandates, some in response to ex/al concerns, some to be achieved through a competitive grants process. In addition, the 1977 law authorized a doubling of institutional funding. It was supported by both the industry and the ex/al coalitions.

The promise of this consensus did not materialize. The administration did not provide increased institutional funding at the level anticipated. So the state experiment stations and extension services organized their own lobby, in the interest of increasing their share of the budget at the expense of proposed competitive grants funding. This lobby found its principal support within the House Appropriations Subcommittee, the agency least conciliatory toward the ex/al agendas. Meanwhile, agricultural research leaders protested the new federal influence, but voiced no umbrage at the fact of increased industry influence, coming through commodity checkoff mechanisms. Under these checkoff mechanisms, funds obtained by a tax upon a commodity are dispensed by a board associated with a major commodity organization. Thus the government's taxing power is used, but there is no obligation to justify expenditures in legislative hearings and debates, where ex/al groups might generate public concern.

The commodity interest groups, holding the power to give or withhold checkoff funds and other support, began again to use their political leverage to censor ex/al research agendas which conflicted with their interests, as in the case of nutrition research. The long-neglected subject of human nutrition had regained visibility in the past two decades during the controversy over the food stamp programs. In the 1977 farm bill Congress authorized substantial funding for a new nutrition center, and also for competitive grants to study human nutrition. One grant was awarded to a study of the basis of vegetarian food choice, which sought to understand how taste for foods develops. The grant won Sen. William Proxmire's derisive and flippant "Golden Fleece Award." The granting of this award, according to two former chiefs of the office in charge of competitive grants (CRGO), "somehow prompted a number of lobbyists for meat-producing groups to assume that the CRGO was endorsing vegetarianism. Congress specified that no more funds would be spent on research on the social and behavioral aspects of human nutrition."[1] In addition, funding for human nutrition research grants was cut substantially.

In view of the history of agricultural research institutions, one is tempted to react with a "déjà vu" to the developments which have followed since the 1977 law: the return of administrators to reliance upon industry support, the reimposition of clientele censorship, and the abolition of the Science and Education Administration. But in fact much happened which has not been reversed. The Science and Education Administration was more than just another attempt to coordinate agricultural research from the federal level. This effort to merge industry and externality interests did improve communication and understanding on both sides, and did generate large-scope undertakings such as research on integrated pest management. It did, at least temporarily, improve the image of both the USDA and the agricultural research institutions in the eyes of other bureaucracies, concerned publics, and the larger science community.

However, the abolition of SEA undoubtedly signals retrenchment. Some research chiefs now feel free to abandon new research initiatives, such as that on organic farming, and to return to a more comfortable relationship with industry clienteles. Public interest groups, too, are returning to the comfortable stance of critic, and advocate of governmental regulation. A formidable externalities coalition is preparing to attack the confinement system for meat and egg production, which public researchers developed and to which the industry is increasingly committed. This disparate ex/al bloc includes vegetarians who oppose the use of animal products, animal defenders who believe confinement is cruel, and family farm advocates who know that confinement systems are speeding the shift to large-scale agriculture.

If it seems, from these developments, that research politics is returning to the politics of confrontation between ex/al and industry, it should be noted that the participants are at least more knowledgeable. Ex/al groups realize that the family farm has already been grievously undermined, not only by technology, but also by economic and cultural influences and public policy. Industry groups know that large-scale agriculture too is being threatened by, among other things, the high cost of energy and transportation, competition for irrigation water, salinization, and resistant insects. It is increasingly clear to all observers that agricultural production is being expanded now by mining rather than husbanding our agricultural resources. Concurrently, a large public has become interested and informed about agricultural issues. Ex/al groups may find this public easy to reach but no longer likely to be stampeded by sensational charges.

In this environment, various elites which have gained access to research policy making may continue to perform important tasks. Research leaders can continue to design and seek support for research strategies that achieve breakthroughs in basic knowledge. Leaders in foundations and others who created the international research centers can continue to find ways to adapt technology to the needs of developing countries. Ex/al leaders, communicating with concerned publics, may continue to judge agricultural research from a nonindustry perspective. A research politics fed by such contrasting visions may be preferable to one mainly guided only by the pork barrel impulse and by producer anxieties over plant or animal diseases.

Larger visions have always influenced agricultural research, sometimes behind the scenes. The founders of research institutions shared the nineteenth century aspiration, achieved beyond Grandfather's dreams, of removing the dawn-to-dusk drudgery from rural life. The goals of leisure and food abundance, prominent in nineteenth century utopian literature, seem always to have been important in research policy even though not always well articulated. It is interesting that, as leisure time developed, the extension services helped farm families use it, through such work-associated activities as cake decorating and the showing of animals. Extension has been criticized for its leisure education on the grounds that such activity is frivolous, as well as on the grounds that leisure is an option for only a favored few of the many persons displaced from agriculture. Still, Extension was exemplary in recognizing that one challenge met had given rise to another, and that one too should be addressed.

Having praised the success of earlier generations in increasing food abundance and reducing farm drudgery, we should take their example in setting goals appropriate to the challenges of the approaching century. Debate over the future research agenda has produced a number of goals to be added, including these: to distribute generously the benefits of leisure and food abundance, to preserve opportunities for citizens to engage in productive agriculture, and to sustain our productivity by shepherding the resources on which it depends and by protecting the natural environment in which it flourishes. Those who do agricultural research, and those who decide which research should be done, should continue their efforts to reconcile and support these goals within the agenda of agricultural research.

NOTES

CHAPTER ONE

1. U.S. Department of Agriculture, Economic Research Service, *Economic Tables*, ERS 559, 1977.
2. Thomas M. Arndt and Vernon W. Ruttan, "Valuing the Productivity of Agricultural Research: Problems and Issues," in Thomas M. Arndt, Dana G. Dalyrmple, and Vernon W. Ruttan, eds., *Resource Allocation and Productivity in National and International Agricultural Research*, p. 6.
3. Willis L. Peterson and Joseph C. Fitzharris, "Organization and Productivity of the Federal-State Research System in the United States," in Arndt, Dalrymple, and Ruttan, *Agricultural Research*, p. 77.
4. Kenneth John Meier, "The Agricultural Research Service and Its Clientele: The Politics of Food Research" (Paper delivered at the Annual Meeting of the Southwestern Political Science Association, Dallas, Texas, March 30–April 2, 1977), p. 18.
5. Vernon W. Ruttan, "The Objectives of Agricultural Research: Some Hypotheses," draft article, November 28, 1978.
6. Jim Hightower, *Hard Tomatoes, Hard Times: The Failure of the Land Grant College Complex* (Agribusiness Accountability Project, 1972).
7. Jim Hightower, *Eat Your Heart Out: Food Profiteering in America* (New York: Crown Publishers, 1975).
8. Hightower, *Hard Tomatoes, Hard Times*, p. 110.
9. Ibid., p. 111.
10. "Public Interest Groups," *National Journal Reports* 6 (February 23, 1974): 274–75.
11. U.S., Congress, House, Committee on Science and Technology, *Special Oversight Hearings on Agricultural Research and Development*, 94th Cong., 1st sess., 1975, pp. 6–37.

CHAPTER TWO

1. For a description of Hatch Act supporters, see Vernon Carstenson, "The Genesis of an Agricultural Experiment Station," *Agricultural History* 34 (January, 1960): 13–20. Gladys Baker, Wayne D. Rasmussen, Vivian Wiser, and Jane M. Porter refer to this group as being influential in the establishment of the Department of Agriculture, in *Century of Service: The First One Hundred Years of the United States Department of Agriculture*, pp. 2–12.

2. Wayne D. Rasmussen and Gladys L. Baker, *The Department of Agriculture,* pp. 14–15.

3. Carroll W. Purcell, Jr., "The Administration of Science in the Department of Agriculture, 1933–1940," *Agricultural History* 42 (July, 1968): 235; Baker et al., *Century of Service,* p. 227; and Gladys Baker, "Notes on Implementing the Sciences—A Case: Organizing Scientific Research in the Department of Agriculture," mimeographed paper prepared for Assistant Secretary of Agriculture Joseph Robertson (Washington, D.C., undated, c. 1964), pp. 17–19.

4. Baker, "Notes on Implementing the Sciences," p. 5.

5. Ibid.

6. Ibid., pp. 9–10.

7. Ibid.

8. Charles Hardin, *Freedom in Agricultural Education,* pp. 69–70.

9. Ibid., pp. 102–6.

10. Ibid., pp. 87–88.

11. Jane Porter, "Experiment Stations in the South, 1877–1940," *Agricultural History* 53 (January, 1979): 85–86.

12. The recommendations are found in the following publications: General Accounting Office, *Management of Agricultural Research: Need and Opportunities for Improvement* (Report to the Joint Economic Committee by the Comptroller General of the United States, August 23, 1977); U.S., Congress, House, Committee on Science and Technology, *Special Oversight Review of Agricultural Research and Development,* 94th Cong., 2nd sess., 1976; National Academy of Sciences, *World Food and Nutrition Study* (1977); U.S., Congress, Office of Technology Assessment, *Organizing and Financing Basic Research to Increase Food Production* (June, 1977); National Research Council, *Report on Research Advisory to the U.S. Department of Agriculture* (1972).

13. Hardin, *Freedom in Agricultural Education,* pp. 26–27.

14. H. C. Knoblauch, E. M. Law et al., *State Agricultural Experiment Stations: A History of Research Policy and Procedure,* p. 116.

15. Vivian Wiser and Douglas Bowers, "Research and Its Coordination in USDA: A Historical Approach," mimeographed paper (Washington, D.C.: Department of Agriculture, May, 1979), p. 65.

16. Interview with Thomas Ronningen, associate administrator, Cooperative State Research Service, 1977.

17. Knoblauch, Law et al., *State Agricultural Experiment Stations,* p. 94.

18. Ibid., p. 117.

19. John R. Steelman, "The President's Scientific Research Board, Administration for Research: A Report to the President," *Science and Public Policy* 3

(October, 1947): 86 (quoted in Baker, "Notes on Implementing the Sciences," p. 21).

20. Roy Lovvorn, former administrator of the Cooperative State Research Service, in a speech before the Senior Staff Conference of the Agricultural Research Service, October 3, 1977.

21. Testimony of Robert Long and Orville Bentley, cochairmen of ARPAC. U.S., Congress, House, Committee on Science and Technology, *Hearings on Agricultural Research and Development,* 94th Cong., 1st sess., 1975, pp. 45–46.

CHAPTER THREE

1. U.S. Department of Agriculture, Agricultural Research Service, *The Beltsville Agricultural Research Center.*

2. Karl Quisenberry, "The Dry Land Stations: Their Mission and Their Men," *Agricultural History* 51 (January, 1977): 218–28.

3. U.S. Department of Agriculture, Cooperative States Research Service, *Professional Workers in State Agricultural Experiment Stations and Other Cooperating State Institutions,* Agriculture Handbook no. 305, 1976.

4. Isao Fujimoto and Emmett Fiske, "What Research Gets Done at a Land Grant College: Internal Factors at Work," (unpublished paper, undated, c. 1974).

5. Ibid.

6. Heather Johnston Nicholson, "The Politics of Research of Southern Corn Leaf Blight" (Paper delivered at the Annual Meeting of the Southwestern Political Science Association, Dallas, Texas, March 30–April 2, 1977).

7. Discussed by Stevan Dedijer, "The Future of Research Priorities," in L. W. Bass and B. S. Old, *Formulation of Research Policies,* AAAS Publication no. 87 (Washington, D.C., 1967), pp. 141–62.

8. Interview with Charles E. Lewis, National Program Staff, Agricultural Research Service, July 20, 1977.

9. Harold Breimyer, *Individual Freedom and the Economic Organization of Agriculture* (Urbana, Ill.: University of Illinois Press, 1965).

10. U.S. Department of Agriculture, Agricultural Research Service, *An Evaluation of Research on Lymphoid Leukosis and Marek's Disease* (June, 1975), table 12.

11. Ibid., table 13.

12. Ibid., table 15.

13. Interview with Graham Purchase, National Program Staff, Agricultural Research Service, 1977.

14. Lawrence Busch, William B. Lacy, and Caroline Sachs, *Research Policy and*

Process in the Agricultural Sciences: Some Results of a National Study, RS-66 (July, 1980), Department of Sociology, Kentucky State Agricultural Experiment Station, Lexington, Kentucky.

15. U.S. Department of Agriculture, Office of Personnel, and Personnel Division of the Agricultural Research Service, *Profile of Scientists in Research Activities of the Agricultural Research Service* (Joint Staff Report, March, 1970), pp. 8–9.

16. Don K. Price, *The Scientific Estate,* p. 64.

17. Even on current occupational status scales, farming ranks relatively low; for example, Otis Dudley Duncan, "Decile Prestige Scale of Occupations," in *Codebook,* Center for Political Studies, Institute for Social Research, 1972 American National Election Study, pp. 656–96.

18. Interview by Jan Bower with Prof. Henry Gilman, on the occasion of the centennial of the Iowa State Graduate School, 1976.

19. Interview with Susan DeMarco, 1975.

20. Interview with James B. Kendrick, Jr., vice-president for Agricultural Research, Director of Cooperative Extension, and Director of the Station (University-wide), Berkeley, August 5, 1977.

21. From data compiled in 1977, by Louis Thompson, associate dean of the College of Agriculture, Iowa State University.

22. *Washington Post,* August 1, 1977, p. 1.

23. Ibid.

24. Enrollment "in agriculture and related programs in 70-member institutions of NASULGC [the National Association of State Universities and Land Grant Colleges] increased approximately 160 percent from 1963 to 1976." From the testimony of C. B. Browning, dean of Resident Instruction, Institute of Food and Agricultural Services, University of Florida, in U.S., Congress, House, Committee on Agriculture, *Hearings on the National Agricultural Research Policy Act of 1976,* 94th Cong., 1st sess., 1976, p. 104.

25. *Washington Post,* August 1, 1977, p. 1.

26. *Des Moines Register,* June 12, 1977, p. 1F.

27. Interview with Harold Heady, associate dean, California Experiment Station, Berkeley, August 5, 1977.

28. Interview with John Pesek, chairman, Department of Agronomy, Iowa State University, 1977.

29. One study of agricultural research workers found that the quantity of scientific output tends to *increase* with age, but the quality of output, measured by most-quoted contributions, was highest for workers in the early stages of their careers. In J. Lewin, "A Quantitative and Qualitative Case-

Study Analysis of Scientific Productivity in Agricultural Research," *Israel Journal of Agricultural Research* 22 (September, 1972): 137.

30. Interview with James B. Kendrick, August 5, 1977.

CHAPTER FOUR

1. Arnon, *Organization and Administration of Agricultural Research,* p. 243.
2. Donald C. Pelz and Frank M. Andrews, *Scientists in Organizations: Productive Climates for Research and Development,* pp. 124–25.
3. Information about the Bradfield-Extension case was obtained from interviews with knowledgeable persons and from three news articles: Bob Schildgren, "The Nightmare World of a Berkeley Professor," *Berkeley Barb,* February 17–23, 1978; Peter Schrag, "The Hounding of Robert Bradfield," *Sacramento Bee,* March 24, 1978; and the *University Guardian* (American Federation of Teachers newspaper), January–February, 1978.
4. Interview with Wise Burroughs, March, 1977.
5. Alex F. McCalla, "Politics of the Agricultural Research Establishment," in Don F. Hadwiger and William P. Browne, eds., *The New Politics of Food,* p. 88.
6. Tom Pearse, "New Varieties Spell Success Story," *Wheat Growers News* (official publication of the Idaho Wheat Growers Association), July, 1976, p. 4.
7. McCalla, "Agricultural Research Establishment," p. 87.
8. Don K. Price, *The Scientific Estate,* p. 270.
9. Arnon, *Organization of Agricultural Research,* p. 166.
10. Interview with Randy Altman, August, 1977.
11. Testimony of Glenn Pound, U.S., Congress, House, Committee on Science and Technology, *Agricultural Research and Development: Special Oversight Hearings,* 94th Cong., 1st sess., 1975, pt. 2, pp. 431–32.
12. Interview with Wise Burroughs, May 17, 1976.
13. Quoted in Spencer Klaw, *The New Brahmins: Scientific Life in America,* p. 83.
14. Price, *The Scientific Estate,* p. 214.
15. Speech to land-grant administrators, 1974.
16. Interview with Dale Anderson, senior specialist, Agricultural Research Service, October 11, 1977.
17. Such losses remain at 20 percent of all food produced in the United States, as reported in, U.S. General Accounting Office, *Food Waste: An Opportunity to Improve Resource Use: Report to the Congress by the Comptroller General of the United States,* Report no. CED–77–118, 1977, p. i.
18. Dale L. Anderson (National Planning Staff, Agricultural Research Service,

USDA), "White (or Slightly Gray) Paper in Defense of 'Agricultural Marketing' Research," mimeographed, January 6, 1977.

19. Indiana State University, School of Business, Bureau of Business Research, and U.S. Department of Agriculture, Agricultural Research Service, Transportation and Facilities Research Division, "Computerized Checkout Systems for Retail Food Stores," *Management Information Bulletin*, no. 3 (School of Business, Indiana State University, April, 1971), p. 1.

20. *Consumer Reports*, May, 1974, pp. 364–65.

21. *U.S. News and World Report*, December 30, 1974, pp. 56–58.

22. Frances Serra, "A Lobbyist for Consumers," *New York Times*, October 31, 1976, sect. 3, p. 7.

23. Bernice T. Eiduson, *Scientists: Their Psychological World* (New York: Basic Books, 1962), p. 124.

CHAPTER FIVE

1. Michael Polanyi, "The Republic of Science: Its Political and Economic Theory," in Edward A. Shils, ed., *Criteria for Scientific Development: Public Policy and National Goals*, p. 3.

2. Thomas M. Arndt and Vernon W. Ruttan, "Valuing the Productivity of Agricultural Research: Problems and Issues," in Thomas M. Arndt, Dana G. Dalrymple, and Vernon W. Ruttan, eds., *Resource Allocation and Productivity in National and International Agricultural Research*, p. 18.

3. Ibid., p. 8.

4. Indeed, Hayami and Ruttan are among those who have indicated apprehension that, in developing countries, inappropriate technology might too quickly displace rural labor. In Yujito Hayami and Vernon W. Ruttan, *Agricultural Development: An International Perspective*, p. 293.

5. Donald R. Kaldor, "Social Returns to Research and the Objectives of Public Research," in Walter L. Fishel, ed., *Resource Allocation in Agricultural Research*, p. 63.

6. See Arndt, Dalrymple, and Ruttan, *Agricultural Research*, p. 5.

7. Kenneth E. Boulding, "Agricultural Organizations and Policies: A Personal Evaluation," in Iowa State University Center for Agricultural and Economic Development, *Farm Goals in Conflict*, p. 161.

8. National Academy of Sciences, National Research Council, *Report of the Committee on Research Advisory to the U.S. Dept. of Agriculture*, 1972.

9. National Academy of Sciences, National Research Council, *Report of the Committee on Research*, pp. 10–12.

10. Jim Hightower, *Hard Tomatoes, Hard Times: The Failure of the Land Grant College Complex*, especially pp. 145–49.

11. National Academy of Sciences, *Committee on Agricultural Production Efficiency*, 1975; National Academy of Sciences, Board on Agricultural and Renewable Resources, *Enhancement of Food Production in the United States*, 1975.

12. National Academy of Sciences, National Research Council, *World Food and Nutrition Study: Potential Contributions of Research*, 1977. See also vol. 5 of the supporting papers.

13. U.S., Congress, Office of Technology Assessment, *Organizing and Financing Basic Research to Increase Food Production* (Washington, D.C., 1977).

14. U.S., Congress, House, Committee on Science and Technology, *Special Oversight. Review of Agricultural Research and Development*, 94th Cong., 2nd sess., 1976, p. 4.

15. Ibid., pp. 1, 3.

16. Jim Hightower, *Hard Tomatoes Hard Times*, p. 144.

17. U.S. Department of Agriculture and National Association of State Universities and Land Grant Colleges, *A National Program for Agricultural Research* (1966).

18. U.S., Congress, General Accounting Office, *Management of Agricultural Research: Need and Opportunities for Improvement: Report to the Joint Economic Committee by the Comptroller General of the United States*, Report no. CED–77–121, 1977.

19. Interview with Waldemar Klassen, National Program Staff, ARS, October 28, 1977.

20. For a discussion of methods which have been used to obtain group decisions, see C. Richard Shumway, "Models and Methods Used to Allocate Resources in Agricultural Research: A Critical Review," in Arndt, Dalrymple, and Ruttan, *Agricultural Research*, pp. 436–57.

21. Arndt and Ruttan, "Valuing Productivity," p. 7.

22. Reed Hertford and Andrew Schmitz, "Measuring Economic Returns to Agricultural Research," in Arndt, Dalrymple, and Ruttan, *Agricultural Research*, p. 162.

23. Ned Bayley, "Research Resource Allocation in the Department of Agriculture," in Fishel, *Resource Allocation*, p. 231.

24. Arndt and Ruttan, "Valuing Productivity," p. 20.

25. Testimony of Dr. Thomas Ronningen, U.S., Congress, House, Committee on Science and Technology, *Agricultural Research and Development: Special Oversight Hearings*, 94th Cong., 1st sess., 1975, pt. 2, p. 150.

26. Interview with Charles Lewis, National Planning Staff, Agricultural Research Service, July 20, 1977.

27. Interview with Edward Jaenke, Jaenke and Associates, June, 1979.

CHAPTER SIX

1. The number of general organizations included in table 6.1, however, needs explanation to avoid exaggeration: several groups were state affiliates of a general organization.

2. Don Paarlberg reviews these accusations in *Farm and Food Policy: Issues of the 1980s* (Lincoln: University of Nebraska Press, 1980).

3. H.O.Carter and Warren E.Johnston, "Some Forces Affecting the Organization and Control of American Agriculture," *American Journal of Agricultural Economics* 60 (1978): 742–43.

4. Estimate by Lawrence Busch, associate professor of Sociology, University of Kentucky, at a Workshop on Agricultural Research in the Private and Public Sectors: Goals and Priorities, Lexington, Kentucky, April 14, 1981.

5. Agricultural Research Institute, *A Survey of U.S. Agricultural Research by Private Industry* (1977), table 2.

6. Ibid., p.11.

7. Ibid., table 5.

8. Interview with Charles Boothby, president, National Association of Soil Conservation Districts, June 20, 1979.

9. Karen A. Schwartz, "Analysis of Factors Influencing Selection of Research Topics in the Kansas State Agricultural Experiment Station" (Master's thesis, Kansas State University, 1978), p.68.

10. Kenneth John Meier, "The Agricultural Research Service and Its Clientele: The Politics of Food Research" (Paper delivered at the Annual Meeting of the Southwestern Political Science Association, Dallas, Texas, March 30–April 2, 1977), p.18.

11. Kenneth John Meier, "Building Bureaucratic Coalitions: Client Representation in USDA Bureaus," in Don F. Hadwiger and William P. Browne, eds., *The New Politics of Food*, p.67.

12. Ibid., p.68.

13. Isao Fujimoto and William Kopper, "Outside Influences in What Research Gets Done at a Large Land Grant School: Impact of Marketing Orders" (Paper delivered at the Annual Meeting of the Rural Sociological Society, San Francisco, California, August 21–24, 1975).

14. Jim Hightower, *Hard Tomatoes, Hard Times: The Failure of the Land Grant College Complex*, p.11.

15. See, for example, the publications: National Association of State Universities and Land Grant Colleges, Office of Research and Information, *People to People*, 1973; U.S. Department of Agriculture, and National Association of State Universities and Land Grant Colleges, *A People and a Spirit*, 1968.

16. Clyde Manwell and C.M.Ann Baker, *Molecular Biology and the Origin of Species* (Seattle: University of Washington Press, 1970), chap. 12.
17. This comment is based on preliminary findings of a forthcoming report by the Center for Rural Affairs, Walthill, Nebraska, on animal science research decision making at the University of Nebraska. The author served on an advisory committee for this study.
18. *Agricultural Research Institute* (Pamphlet of the Agricultural Research Institute, Joseph Henry Building, Room 835, 2100 Pennsylvania Ave., N.W., Washington, D.C. 20037).
19. Paul J. Feldstein, "The Political Environment of Regulation," in Arthur Levin, ed., *Regulating Health Care* (New York: Academy of Political Science, 1981), p.16.

CHAPTER SEVEN

1. Kenneth John Meier, "The Agricultural Research Service and Its Clientele: The Politics of Food Research" (Paper delivered at the Annual Meeting of the Southwestern Political Science Association, Dallas, Texas, March 30–April 2, 1977), p.6.
2. U.S., Congress, Senate, Committee on Appropriations, *Hearings on Agriculture and Related Agencies Appropriations Bill, 1978*, 95th Cong., 1st sess., 1977, pt. 3, p.1061.
3. Richard F. Fenno, Jr., *The Power of the Purse: Appropriations Politics in Congress* (Boston: Little, Brown and Co., 1966), p.562.
4. Ibid., p.141.
5. Ibid.
6. Biographical information is from the *Congressional Directory*.
7. Heather Johnston Nicholson, "The Politics of Research on Southern Corn Leaf Blight" (Paper delivered at the Annual Meeting of the Southwestern Political Science Association, Dallas, Texas, March 30–April 2, 1977), pp.4–5.
8. See, for example, U.S., Congress, House, Committee on Appropriations, *Hearings before the Subcommittee on Agricultural-Environmental and Consumer Protection Agencies Appropriations*, 92nd Cong., 2nd sess., 1972, pt. 1, p.220.
9. U.S., Congress, House, Committee on Appropriations, *Hearings before the Subcommittee on Department of Agriculture and Related Agencies Appropriations*, 87th Cong., 2nd sess., February 7, 1962, pp.78–79.
10. Oleomargarine research was once a disruptive issue at Iowa State College. For a review of the controversy see Muriel Ann Weir, "Pamphlet No. 5 and

211

the Freedom to Publish at Iowa State College" (Master's thesis, Iowa State University, 1976), especially p. 298.

11. U.S., Congress, House, Committee on Appropriations, *Hearings on Agriculture Appropriations Bill for 1947*, 79th Cong., 2nd sess., 1946, p. 238.

12. U.S., Congress, House, Committee on Appropriations, *Hearings before the Subcommittee on Department of Agriculture and Related Agencies Appropriations*, 87th Cong., 1st sess., March 27, 1961, pt. 1, p. 70.

13. The centennial history is Gladys L. Baker, Wayne D. Rasmussen, Vivian Wiser, and Jane M. Porter, *Century of Service: The First One Hundred Years of the Department of Agriculture*.

14. McHarg's speech was published in a USDA pamphlet. The Appropriations Subcommittee's discussion of the McHarg lecture, from which subsequent quotes are drawn, is in U.S., Congress, House, Committee on Appropriations, *Hearings before the Subcommittee on Agriculture-Environmental and Consumer Agency Appropriations*, 92nd Cong., 2nd sess., February 22, 1972, pt. 1, pp. 273–375.

15. U.S., Congress, House, Committee on Appropriations, *Hearings before the Subcommittee on Agriculture-Environmental and Consumer Protection Agencies Appropriations*, 92nd Cong., 2nd sess., February 28, 29, March 2, 6, 1972, pt. 2, p. 394.

16. Ibid., March 2, 1976, pt. 3, pp. 85–87. The reference is to the administrator of the Agricultural Research Service, T. W. Edminister, who is an agricultural engineer.

17. Statements quoted from the conference committee are from notes taken by the author, who attended the committee meeting.

18. Fenno, *The Power of the Purse*, pp. 141, 508, 541, 573–76.

19. Aaron Wildavsky, *The Politics of the Budgetary Process*, 2nd ed. (Boston: Little, Brown and Co., 1974), p. 53.

20. Some agency and bureau budgets, of course, diverged from the predominant direction in any given year.

CHAPTER EIGHT

1. Rachel Carson, *Silent Spring* (Boston: Houghton Mifflin Co., 1962).

2. Jamie L. Whitten, *That We May Live* (Princeton, N.J.: D. Van Nostrand, 1966).

3. U.S., Congress, House, Committee on Agriculture, *Hearings on the Federal Insecticide, Fungicide, and Rodenticide Act*, 69th Cong., 2nd sess., 1946, p. 14.

4. Mel Dubnick, "From Facilitation to Control: Changes in the Regulatory Relationship between Government and Agriculture" (Paper presented at the Fourth Annual Hendricms Symposium, University of Nebraska, Lin-

coln, April 5–6, 1979): "Until the middle 1960's, regulations associated with the use of pesticides by farmers were so facilitative that they should be better termed 'promotional.' "

5. James C. Horsfall, "A Socio-Economic Evaluation," in C.O.Chichester, ed., *Research in Pesticides* (New York: Academic Press, 1965), pp. 3–16.

6. Carson, *Silent Spring*, p. 7.

7. Robert Van den Bosch, *The Pesticides Conspiracy* (Garden City, N.Y.: Doubleday, 1978).

8. Berry recognizes that many groups feel that they are seeking the public interest; for purposes of his study, public interest groups are defined as those which seek a collective good, the achievement of which will not selectively and materially benefit the membership or activists of the organization. Jeffrey M. Berry, *Lobbying for the People*, p. 7. Group characteristics described in the text are summarized from Berry's book.

9. Lester Milbrath, *The Washington Lobbyists* (Chicago: Rand McNally, 1963).

10. Berry, *Lobbying for the People*, p. 97.

11. Ibid., p. 268.

12. Summaries of recommendations from both the president's committee study and the committee's hearings are found in, U.S., Congress, Senate, Committee on Government Operations, *Interagency Environmental Hazards Coordination*, Report no. 1379, 89th Cong., 2nd sess., 1966.

13. U.S., Congress, House, Committee on Government Operations, *Hearings on Deficiencies in Administration of Federal Insecticide, Fungicide, and Rodenticide Act*, 91st Cong., 1st sess., 1969.

14. J.R.M.Innes et al., "Bioassay of Pesticides and Industrial Chemicals for Tumorigenicity in Mice: A Preliminary Note," *Journal of the National Cancer Institute* 42 (June, 1969): 1101–14.

15. U.S., Congress, House, Committee on Appropriations, *Hearings on Department of Agriculture Appropriations for 1966*, 89th Cong., 1st sess., 1965, pt. 1, "Effects, Uses, Control, and Research of Agricultural Pesticides," Report by the Surveys and Investigations Staff, pp. 168ff.

16. Ibid., p. 169.

17. U.S., Congress, Senate, Committee on Interstate and Foreign Commerce, *Hearings on Fish and Wildlife Legislation*, 85th Cong., 2nd sess., 1958, p. 17.

18. P.L. 592, 85th Cong., 2nd sess., 1958.

19. *Congress and the Nation, 1945–1964* (*Congressional Quarterly*, 1965).

20. U.S., Congress, House, Committee on Appropriations, *Hearings on Agriculture, Rural Development, and Related Agencies Appropriations for 1980*, 96th Cong., 1st sess., 1979, pt. 3, p. 186.

21. U.S. Department of Health, Education, and Welfare (HEW), *Report of the*

Secretary's Commission on Pesticides and their Relationship to Environmental Health (1969).

22. In 1974, U.S. Agricultural Research Service funds for pest control were allocated as follows: 29 percent for basic research; 44 percent for alternatives on pesticide development; and 8 percent on metabolism residues (Source: Interview with M.Ouye, ARS).

23. U.S. Department of Health, Education, and Welfare, *Report of Secretary's Commission on Pesticides.*

24. U.S., Congress, Senate, Committee on Government Operations, *Interagency Environmental Hazards Coordination,* 89th Cong., 2nd sess., 1966, table 3.

25. See, for example, Council of Agricultural Science and Technology, "Chlordane and Heptachlor," Report no.47, October 3, 1975. This seventy-one–page report was prepared for EPA hearings on the suspension of use of these two insecticides.

26. "Cast-Industry Tie Raises Credibility Concerns," *BioScience* 29 (January, 1979): 58–59.

27. *Des Moines Register,* July 31, 1977.

28. See U.S., Congress, Senate, Committee on Appropriations, *Hearings on Agriculture, Rural Development, and Related Agencies Appropriations for 1980,* 96th Cong., 1st sess., 1979, pt. 1, p.150.

CHAPTER NINE

1. Keith O. Campbell, *Food for the Future* (Lincoln: University of Nebraska Press, 1979).

2. Ibid., p.42.

3. Ibid., p.44.

4. Ibid., p.58.

5. See Robert E. Evenson, Paul E. Waggoner, and Vernon W. Ruttan, "Economic Benefits from Research: An Example from Agriculture," *Science* 205 (September 14, 1979): 1101–07. Also, Theodore Schultz, "The Economics of Research and Agricultural Productivity" (Paper delivered at the Seminar on Socio-Economic Aspects of Agricultural Research in Developing Countries, May 7–11, 1979, Santiago, Chile), pp.3–4.

6. Schultz, "Economics of Research," p.4.

7. Frances Moore Lappé and Joseph Collins, *Food First: Beyond the Myth of Scarcity* (Boston: Houghton Mifflin Co., 1977). See, for example, pp. 143–55.

8. Campbell, *Food for the Future,* p.20.

9. Willard Cochrane, *U.S. Trade Policy and Agricultural Exports* (Ames, Iowa: Iowa State Press, 1972), pp. 196–97.
10. Geoffrey Paige, *Agrarian Revolution: Social Movements and Export Agriculture in the Underdeveloped World*, pp. 14–16, 46.
11. Donald Puchala and Jane Staveley, "The Political Economy of Taiwanese Agricultural Development," in Raymond F. Hopkins, Donald J. Puchala, and Ross B. Talbot, eds., *Food, Politics and Agricultural Development: Case Studies in the Public Policy of Rural Modernization* (Boulder, Colo.: Westview Press, 1979), pp. 107–32.
12. See Barry Commoner, *The Poverty of Power: Energy and Economic Crisis*, especially pp. 155–75; and the testimony of Barry Commoner in, U.S., Congress, House, Committee on Science and Technology, *Agricultural Research and Development: Special Oversight Hearings*, 94th Cong., 1st sess., 1975, pt. 2, pp. 513–49.
13. Garth Youngberg, "The Alternative Agricultural Movement: Its Ideology, Its Politics, and Its Prospects" (Paper delivered at the Annual Meeting of the Southwestern Political Science Association, Dallas, Texas, March 30–April 2, 1977).
14. E. F. Schumacher, *Small Is Beautiful: Economics as if People Mattered*, p. 14.
15. Youngberg, "The Alternative Agricultural Movement."
16. Lappé and Collins, *Food First*, especially pp. 111–18.
17. Wendell Berry, *The Unsettling of American Agriculture*, p. 95 (Berry is quoting Ivan Illich).
18. Agricultural Research Policy Advisory Committee, *Research to Meet U.S. and World Food Needs*, Report of Working Conference, Kansas City, Missouri, July 9–11, 1975.
19. U.S., Congress, House, Committee on Appropriations, *Hearings on Agriculture, Rural Development, and Related Agencies Appropriations for 1981*, 96th Cong., 1st sess., 1980, pp. 10–11.

CHAPTER TEN

1. Charles Jones, "Representation in Congress: The Case of the House Agriculture Committee," *American Political Science Review* 55 (June, 1961): 358–67.
2. It was so characterized by Leonard D. White, in *The Republican Era: 1869–1901* (New York: Macmillan Co., 1958), chap. 11.
3. Theodore J. Lowi, *The End of Liberalism* (New York: W. W. Norton, 1969), p. 110.

4. Don F. Hadwiger, "The Freeman Administration and the Poor," *Agricultural History* 45 (January, 1971): 110.
5. Address by Don Paarlberg, director of Agriculture Economics, U.S. Department of Agriculture, at the National Public Policy Conference, Clymer, New York, September 11, 1975.
6. *Des Moines Register,* July 31, 1977, sect. A, p. 1.
7. Speech by Deputy Assistant Secretary James Neilson to Senior Staff Conference, Agricultural Research Service, October 3, 1977.
8. U.S., Congress, House, Committee on Appropriations, *Hearings on Agriculture, Rural Development, and Related Agencies Appropriations for 1981,* 96th Cong., 1st sess., 1980, p. 35.
9. See, for example, U.S., Congress, House, Committee on Appropriations, *Hearings on Agriculture and Related Appropriations for 1978,* 95th Cong., 1st sess., 1977, pp. 1061–64.
10. Discussion by Don Paarlberg, Professor Emeritus, Purdue University, "Food and Agricultural Research Agenda," at the U.S. Dept. of Agriculture Outlook Conference, November 19, 1980.
11. Consultative Group on International Agricultural Research (1976), p. 4–5.

CHAPTER ELEVEN

1. David W. Krogmann and Joe Key, "The Agriculture Grants Program," *Science* 213 (July 10, 1981): 1980.

BIBLIOGRAPHY

Within the large literature on science and public policy, there is recognition that agricultural research is unique in its history, organization, and goals. Not so relevant to agriculture, therefore, are the apprehensions around which many books on science policy have been organized during the last three decades—apprehensions about whether a science establishment is gaining too much political power, or whether government is becoming too prominent in organizing scientific work.

In light of the fact that relationships between agricultural scientists and government firmed up many years ago, agricultural research might have served as a prototype for science policy in other fields. Instead, perhaps because of the presumption that the agricultural science environment is sui generis, science literature devotes relatively little time to an analysis of agricultural science policy.

There is a small body of literature on agricultural research. Within this, some studies explain the history of research institutions. Other studies, usually by economists, seek to understand the impact of agricultural research and to estimate its cost efficiency. These studies may also seek mechanisms or processes by which to formulate and implement "societal goals" in agricultural research.

There is a body of primary sources to be used in the study of agricultural research. Primary sources include the Current Research Information Service, which describes and classifies agricultural research projects; budget documents of the research agencies; annual hearings and reports of the Senate and House subcommittees on agricultural appropriations; some occasional research oversight hearings, including those by the House and Senate agriculture committees and one in 1975 by the House Science and Technology Committee. A number of study groups have submitted written—sometimes voluminous—reports on agricultural research, including groups sponsored by the National Research Council of the National Academy of Science, the General Accounting Office, and the Office of Technology Assessment. Papers of the Agribusiness Accountability Project and the Council for Agricultural and Science Technology, as well as those of some other groups interested in agricultural research, are available in the Special Collections Section of the Iowa State University library. The papers of the secretaries of Agriculture, including recent secretaries, are available through the National Archives.

Bibliography

Since agricultural research policy is in part a product of an industry subsystem and coalition, and recently has been influenced by an externalities / alternatives coalition, reference should be made to the body of literature on these coalitions, some of which appears in this bibliography and prominently in the notes for chapters 6, 8, and 9.

The following selective citations are mainly secondary sources in the three areas noted above: science policy, agricultural science institutions and policy, and political coalitions.

I. SCIENCE POLICY

Burger, Edward J. *Science in the White House: A Political Liability.* Baltimore: The Johns Hopkins University Press, 1980.

Dedijer, Stevan. "The Future of Research Priorities." In L. W. Bass and B. S. Old, *Formulation of Research Policies.* AAAS Publication no. 87, (Washington, D.C., 1967).

Klaw, Spencer. *The New Brahmins: Scientific Life in America.* New York: William Morrow, 1968.

Orlans, Harold, ed. *Science Policy and the University.* Washington, D.C.: Brookings, 1968.

Pelz, Donald C., and Andrews, Frank M. *Scientists in Organizations: Productive Climates for Research and Development.* New York: Wiley, 1966.

Polanyi, Michael. "The Republic of Science: Its Political and Economic Theory." In *Criteria for Scientific Development: Public Policy and National Goals,* edited by Edward A. Shils. Cambridge, Mass.: M.I.T. Press, 1968.

Price, Don K. *The Scientific Estate.* Cambridge: Harvard University Press, 1965.

II. AGRICULTURAL RESEARCH POLICY

Arndt, Thomas M., Dalrymple, Dana G., and Ruttan, Vernon W., eds. *Resource Allocation and Productivity in National and International Agricultural Research.* Minneapolis: University of Minnesota Press, 1977.

Arnon, I. *Organization and Administration of Agricultural Research.* Amsterdam: Elsevier Publishing Company, 1968.

Baker, Gladys L. *The County Agent.* Chicago: University of Chicago Press, 1939.

Baker, Gladys L., Rasmussen, Wayne D., Wiser, Vivian, and Porter, Jane M. *Century of Service: The First One Hundred Years of the United States Department of Agriculture.* Washington, D.C.: Centennial Committee, U.S. Department of Agriculture, Government Printing Office, 1963.

Boulding, Kenneth E. "Agricultural Organizations and Policies: A Personal Evaluation." In Iowa State University Center for Agricultural and Economic

Bibliography

Development, *Farm Goals in Conflict*. Ames, Iowa: Iowa State Press, 1963.

Busch, Lawrence, Lacy, William B., and Sachs, Caroline. *Research Policy and Process in the Agricultural Sciences: Some Results of a National Study*. RS-66 (July, 1980), Department of Sociology, Kentucky State Agricultural Experiment Station, Lexington, Kentucky.

Busch, Lawrence, ed. *Science and Agricultural Development*. Totowa, N.J.: Allanheld, Osmun, 1981.

Carson, Rachel. *Silent Spring*. Boston: Houghton Mifflin Co., 1962.

Carstenson, Vernon. "The Genesis of an Agricultural Experiment Station." *Agricultural History* 34 (January, 1960).

Evenson, Robert E., Waggoner, Paul E., and Ruttan, Vernon W. "Economic Benefits from Research: An Example from Agriculture." *Science* 205 (September 14, 1979).

Fishel, Walter L., ed. *Resource Allocation in Agricultural Research*. Minneapolis: University of Minnesota Press, 1971.

Hadwiger, Don F. "The Freeman Administration and the Poor." *Agricultural History* 45 (January, 1971).

Hardin, Charles. *Freedom in Agricultural Education*. Chicago: University of Chicago Press, 1955.

Hayami, Yujito, and Ruttan, Vernon W. *Agricultural Development: An International Perspective*. Baltimore: The Johns Hopkins University Press, 1971.

Hightower, Jim. *Hard Tomatoes, Hard Times: The Failure of the Land Grant College Complex*. Washington, D.C.: Agribusiness Accountability Project, 1972; and Cambridge, Mass.: Schenkman Publishing Co., 1973.

Kirkendall, Richard S. *Social Scientists and Farm Politics in the Age of Roosevelt*. Columbia, Mo.: University of Missouri Press, 1966.

Knoblauch, H. E., and Law, E. M., et al. *State Agricultural Experiment Stations: A History of Research Policy and Procedure*. USDA, Misc. Publ. 904. Washington, D.C.: Government Printing Office, 1962.

Lappé, Frances Moore, and Collins, Joseph. *Food First: Beyond the Myth of Scarcity*. Boston: Houghton Mifflin Co., 1977.

McCalla, Alex F. "Politics of the Agricultural Research Establishment." In *The New Politics of Food*, edited by Don F. Hadwiger and William P. Browne. Lexington, Mass.: Lexington Books, 1978.

Meier, Kenneth John. "Building Bureaucratic Coalitions: Client Representation in USDA Bureaus." In *The New Politics of Food*, edited by Don F. Hadwiger and William P. Browne. Lexington, Mass.: Lexington Books, 1978.

Moore, Ernest G. *The Agriculture Research Service*. New York: Praeger, 1967.

Moseman, Albert H. *Building Agricultural Research Systems in the Developing Nations*. New York: Agricultural Development Council, 1970.

Bibliography

Porter, Jane. "Experiment Stations in the South, 1877–1940." *Agricultural History* 53 (January, 1979).

Purcell, Carroll W., Jr. "The Administration of Science in the Department of Agriculture, 1933–1940." *Agricultural History* 42 (July, 1968).

Quisenberry, Karl. "The Dry Land Stations: Their Mission and Their Men." *Agricultural History* 51 (January, 1977).

Rasmussen, Wayne D., and Baker, Gladys L. *The Department of Agriculture*. New York: Praeger, 1972.

Rossiter, Margaret W. *The Emergence of Agricultural Science: Justus Liebig and the Americans, 1840–1880*. New Haven: Yale University Press, 1975.

True, Alfred C. *A History of Experimentation and Research in the United States, 1607–1925, including a History of the United States Department of Agriculture*. Washington: Government Printing Office, 1937.

III. COALITION POLITICS

Barton, Weldon V. "Coalition Building in the United States House of Representatives Agricultural Legislation in 1973." In *Cases in Public Policy-Making*, edited by James E. Anderson. New York: Praeger, 1976.

Barton, Weldon V. "Food, Agriculture, and Administrative Adaptation to Political Change." *Public Administration Review* 36 (March–April, 1976).

Berry, Jeffrey M. *Lobbying for the People*. Princeton, N.J.: Princeton University Press, 1977.

Berry, Wendell. *The Unsettling of American Agriculture*. San Francisco: Sierra Club, 1977.

Breimyer, Harold. *Individual Freedom and the Economic Organization of Agriculture*. Urbana, Ill.: University of Illinois Press, 1965.

Campbell, Keith O. *Food for the Future*. Lincoln: University of Nebraska Press, 1979.

Carter, H. O., and Johnston, Warren E. "Some Forces Affecting the Organization and Control of American Agriculture." *American Journal of Agricultural Economics* 60 (December, 1978).

Cochrane, Willard, and Ryan, Mary. *American Farm Policy: 1948–1973*. Minneapolis: University of Minnesota Press, 1976.

Commoner, Barry. *The Poverty of Power: Energy and Economic Crisis*. New York: Alfred A. Knopf, 1976.

Destler, I. M. "United States Food Policy 1972–1976: Reconciling Domestic and International Objectives." *International Organization* 32 (Summer, 1978).

Fraenkel, Richard, Hadwiger, Don F., and Browne, William P., eds. *The Role of U.S. Agriculture in Foreign Policy*. New York: Praeger, 1979.

Bibliography

Gaus, John M., and Wolcott, Leon O. *Public Administration and the United States Department of Agriculture*. Chicago: Public Administration Service, 1940.

Hadwiger, Don F., and Browne, William P., eds. *The New Politics of Food*. Lexington, Mass.: Lexington Books, 1978.

Hadwiger, Don F. "The Old, the New and the Emerging United States Department of Agriculture." *Public Administration Review* 36 (March–April, 1976).

Hajda, Joseph, Michie, Aruna, and Sloan, Thomas, eds. *Political Aspects of World Food Problems*. Manhattan: Agricultural Experiment Station, Kansas State University, 1978.

Jones, Charles. "Representation in Congress: The Case of the House Agriculture Committee." *American Political Science Review* 55 (June, 1961).

Lewis-Beck, Michael. "Agrarian Political Behavior in the United States." *American Journal of Political Science* 21 (August, 1977).

Lowi, Theodore J. "How Farmers Get What They Want." In *Legislative Politics U.S.A.*, edited by Theodore J. Lowi and Randall B. Ripley. 3d ed. Boston: Little, Brown & Co., 1973.

Paarlberg, Don. *Farm and Food Policy: Issues of the 1980s*. Lincoln: University of Nebraska Press, 1980.

Paige, Jeffery M. *Agrarian Revolution: Social Movements and Export Agriculture in the Underdeveloped World*. New York: The Free Press, 1975.

Ruttan, Vernon W. *Agricultural Research Policy*. Minneapolis: University of Minnesota Press, 1981.

Schumacher, E. F. *Small Is Beautiful: Economics as if People Mattered*. London: Harper & Row, 1973.

Talbot, Ross B. "The Three U.S. Food Policies: An Ideological Interpretation." *Food Policy* 2 (February, 1977).

Talbot, Ross B., and Hadwiger, Don F. *The Policy Process in American Agriculture*. San Francisco: Chandler, 1968.

Van den Bosch, Robert. *The Pesticides Conspiracy*. Garden City, N.Y.: Doubleday, 1978.

Whitten, Jamie L. *That We May Live*. Princeton, N.J.: D. Van Nostrand, 1966.

Youngberg, Garth. "U.S. Agriculture in the 1970's: Policy and Politics." In *Economic Regulatory Policies*, edited by James E. Anderson. Lexington, Mass.: Lexington Books, 1976.

INDEX

223